48

TECHNICIAN'S GUIDE TO FIBER OPTICS

THIRD EDITION

TECHNICIAN'S GUIDE TO FIBER OPTICS

THIRD EDITION

DONALD J. STERLING, JR.

DELMAR

THOMSON LEARNING

Africa • Australia • Canada • Denmark • Japan • Mexico • New Zealand • Philippines
Puerto Rico • Singapore • Spain • United Kingdom • United States

Cover Design: Nicole Reamer

Delmar Staff:

Publisher: Alar Elken
Acquisitions Editor: Gregory L. Clayton
Developmental Editor: Amy E. Tucker
Production Manager: Larry Main

Senior Project Editor: Christopher Chien
Art and Design Coordinator: Nicole Reamer
Marketing Coordinator: Paula Collins

COPYRIGHT © 2000
Delmar is a division of Thomson Learning. The Thomson Learning logo is a registered trademark used herein under license.

Printed in the United States of America

For more information contact:

Delmar Publishers
3 Columbia Circle,
Box 15015
Albany, New York USA 12212-5015

International Thomson Publishing Europe
Berkshire House
168-173 High Holborn
London, WC1V 7AA
United Kingdom

Nelson ITP, Australia
102 Dodds Street
South Melbourne, Victoria 3205
Australia

Nelson Canada
1120 Birchmont Road
Scarborough, Ontario
M1K 5G4, Canada

International Thomson Publishing France
Tour Maine-Montparnasse
33, Avenue du Maine
75755 Paris Cedex 15, France

International Thomson Editores
Seneca 53
Colonia Polanco
11560 D.F. Mexico

International Thomson Publishing GmbH
Konigswinterer Straße 418
53227 Bonn
Germany

International Thomson Publishing Asia
60 Albert Street
#15-01 Albert Complex
Singapore 189969

International Thomson Publishing—Japan
Hirakawacho Kyowa Building, 3F
2-2-1 Hirakawa-cho Chiyoda-ku
Tokyo 102 Japan

ITE Spain/Parainfo
Calle Magellanes, 25
28015-Madrid, Espana

5 6 7 8 9 10 XXX 05 04 03 02

Library of Congress Cataloging-in-Publication Data
Sterling, Donald J., 1951–
    Technician's guide to fiber optics / Donald J. Sterling, Jr.—
3rd ed.
        p.   cm.
    1. Fiber optics.   I. Title.
TA1800.S74   1999
621.36'92—dc21   99-33269
                                    CIP

To Lynne and to Megan

CONTENTS

PREFACE TO THE THIRD EDITION

The third edition of *Technician's Guide to Fiber Optics* follows the plan of the previous two editions. It has been updated to reflect the fast-moving technology. New areas of coverage include dense wavelength-division multiplexers and vertical-cavity surface-emitting lasers, while the applications chapter has been broken into two chapters, one dealing with networks and one dealing with telecommunications and CATV. Fiber optics continues to be a fast-moving technology and I have tried to capture the important new trends that have been occurring since the last edition.

The book is intended to be self-contained. Few prerequisites are required, other than some basic familiarity with electronics.

The book is divided into three parts. Part 1, Chapters 1 through 4, attempts to put fiber into perspective as a transmission medium. It not only describes the advantages of fiber over copper counterparts, but it also shows the importance of electronics communications and explains basic concepts such as bits and bytes, analog and digital, and light.

Part 2, Chapters 5 though 12, describes in detail fibers, cables, detectors, connectors and splices, and passive devices such as couplers and multiplexers. Emphasis is given to the theoretical and practical aspects of fibers and connectors in particular, because these are the most different from their electronic counterparts.

Part 3, Chapter 13 through 17, attempts to show how fiber-optics systems are put together. It includes discussions of links (with emphasis on power budgets), installation and hardware, applications, and fiber-optic equipment.

Each chapter contains a summary of important concepts and a set of review questions. A helpful glossary of key terms completes the book.

I have tried to maintain the difficult balance of showing where fiber optics is today by giving examples of the state of the art. I think this is helpful in terms of putting the technology into perspective. But I also realize that the technology changes and that today's state of the art is tomorrow's old technology. Remember that things change and technology marches on. Even as I added new technology, some dated examples have disappeared.

Of the many people who helped make this a better third edition, I would especially like to thank Mike Peppler, Bob Stough, and Clarel Thevenot of AMP; Carl Blesch and Tom Topalian of Lucent Technologies; and Alyson Moore of Siecor.

ACKNOWLEDGMENTS

Many thanks are due to those who reviewed the manuscript and made valuable suggestions for improvement:

Sohail Anwar, Penn State, Altoona, PA
Elias Awad, Wentworth Institute of Technology, Boston, MA
John Carpenter, Sandhills Community College, Pinehurst, NC
Lonnie Lasher, Iowa Central Community College, Fort Dodge, IA
Randy Manning, AMP Incorporated, Middletown, PA
Charles McGlumphy, Belmont Technical Collge, St. Clairesville, OH
Mike Peppler, AMP Incorporated, Middletown, PA
Allen Shotwell, Ivy Tech State College, Terre Haute, IN
Robert Weiler, Capitol College, Laurel, MD
William Zisk, New England Technical Institute, Bristol, CT

I would also like to thank those companies who generously supplied information, artwork, and samples for the book including:

AMP Incorporated
Corning, Inc.
Force, Inc.
Hewlett-Packard Company
IBM
Melcor Corp.
Methode Fiber Optic Products
PIRI
Siecor Corporation
3COM
3M
Lucent Technologies

PART ONE

BACKGROUND

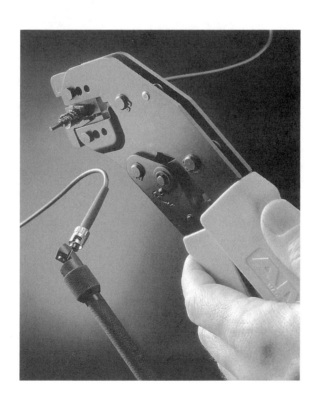

chapter one

THE COMMUNICATIONS REVOLUTION

Fiber optics, as discussed in this book, is simply a method of carrying information from one point to another. An *optical fiber* is a thin strand of glass or plastic that serves as the transmission medium over which the information passes. It thus fills the same basic function as a copper cable carrying a telephone conversation or computer data. Unlike the copper cable, however, the fiber carries light instead of electricity. In doing so, it offers many distinct advantages that make it the transmission medium of choice for applications ranging from telephony and computer communications to automated factories.

The basic fiber-optic system is a link connecting two electronic circuits. Figure 1-1 shows the main parts of a simple link. These parts are as follows:

- *Transmitter,* which converts an electrical signal to an optical signal. The *source,* which is either a light-emitting diode (LED) or a laser diode, does the actual conversion. The source converts electrical current into light. The *drive circuit* changes the electrical signal into the drive current required by the source.
- *Fiber-optic cable,* which is the transmission medium for carrying the light. The cable includes the optical fiber and its protective covering or jacket.
- *Receiver,* which accepts the light and converts it back into an electrical signal. In most cases, the resulting electrical signal is identical to the original signal fed to the transmitter. The two basic parts of the receiver are the *detector,* which converts the optical signal back into an electrical signal, and the *output circuit,* which amplifies, reshapes, and otherwise rebuilds the signal before passing it on.

FIGURE 1-1
Basic fiber-
optic system

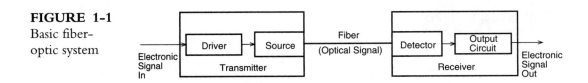

Depending on the application, the transmitter and receiver circuits can be very simple or quite complex. These three parts form the essence of the communications alternative. Other components discussed in this book, such as couplers, multiplexers and optical amplifiers, provide the means for building more complex links and communications networks. The transmitter, fiber, and receiver are the basic elements in every fiber-optic link.

THE HISTORY OF FIBER OPTICS

Using light for communication is not new. Lanterns in Boston's Old North Church sent Paul Revere on his famous ride. Navy signalmen used lamps to communicate between ships with Morse code. For centuries, lighthouses have warned sailors of danger.

Claude Chappe built an optical telegraph in France during the 1790s. Signalmen in a series of towers stretching from Paris to Lille, a distance of 230 km, relayed signals to one another through movable signal arms. Messages could travel end to end in about 15 minutes. In the United States, an optical telegraph linked Boston and the nearby island of Martha's Vineyard. These systems were eventually replaced by electrical telegraphs.

In 1870, the English natural philosopher John Tyndall demonstrated the principle of guiding light through internal reflections. In a presentation before the Royal Society, he showed that light could be bent around a corner as it traveled in a jet of pouring water. Water flowed through a horizontal spout near the bottom of a container, into another container, and through a parabolic path through the air. When Tyndall aimed a beam of light through the spout along the water, his audience saw the light follow a zigzag path inside the curved path of the water. The water trapped and guided the light. An optical fiber works on the same principle of guiding light.

A decade later, Alexander Graham Bell patented the photophone (Figure 1-2), which used unguided light to carry speech. A series of lenses and mirrors threw light onto a flat mirror attached to the mouthpiece. The voice vibrated the mirror, thereby modulating the light hitting it. The receiver used a selenium detector whose resistance varied with the intensity of the light striking it. The voice-modulated sunlight striking the selenium varied the current through the receiver and reproduced the voice. Bell managed to transmit successfully over a distance of 200 m. His invention was novel and interesting, but lacking in the practicality that marked another invention of his: the telephone. Tyndall and Bell, however, demonstrated two important principles that would later be applied to fiber optics: the guiding of light through a transmission medium and the modulation of light to communicate. Practical application, however, required a better transmission medium and better way of modulating light.

Throughout the early twentieth century, scientists made experimental and theoretical investigations into dielectric waveguides, including glass rods. During the 1950s, image-transmitting fibers were developed by Brian O'Brien at the American Optical Company and by Narinder S. Kapany and colleagues at the Imperial College of Science and Technology in London. Such image-carrying fibers today find use in fiberscopes, which are used in medicine to look inside the body. It was Kapany who invented the glass-coated glass rod (which is essentially the structure of today's optical fiber) and, in 1956, coined the term *fiber optics*. The coated glass was an important step, since the outer layer served to keep the light trapped in the inner layer, where it was guided. But the fibers of the 1950s were not yet suited for communications applications.

In 1957, Gordon Gould, a graduate student at Columbia University, described the laser as an intense source of light. Charles Townes and Arthur Schawlow of Bell Laboratories helped to popularize the idea in scientific circles and to spur research into creating a working laser. By 1960, Theodore Maiman of Hughes Laboratories operated the first ruby laser, while Townes demonstrated a helium neon laser. By 1962, lasing was observed in a semiconductor chip, which is the type of laser used in fiber optics. Rather belatedly, Gould was recognized as the father of the laser and was awarded four basic patents in 1988, based on his work in the 1950s.

The importance of the laser as a carrier of information was not lost on communications engineers. A laser had the potential information-carrying capacity of 10,000 times the magnitude of the radio frequencies being used at the time. Despite this potential, the laser was not well suited to open-air, line-of-sight transmission such as was used with microwaves. Fog, smog, rain, and other environmental conditions adversely affect the transmission of laser light. (Just think how fog diminishes the usefulness of a flashlight.) In fact, it is easier to transmit a laser beam from the earth to the moon than from uptown to downtown Manhattan. The laser, then, was a communications source waiting for a suitable transmission medium. Enter the optical fiber.

FIGURE 1-2
Alexander Graham Bell's photophone (Illustration courtesty of AMP Incorporated)

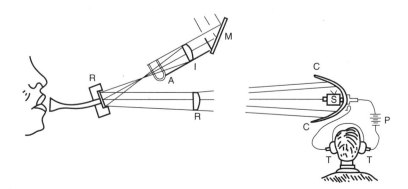

In 1966, Charles Kao and Charles Hockham, of Standard Telecommunications Laboratory in England, published a paper proposing that optical fibers could be used as a transmission medium if their losses could be reduced to 20 dB/km. They speculated that the current high losses of 1000 dB/km were the result of impurities in the glass and not a property of the glass itself. The high losses prevented a fiber from transmitting light over any significant distance. Kao and Hockham reasoned that reducing these impurities would produce low-loss fibers suitable for communications.

By 1970, Robert Maurer, Donald Keck, and Peter Schultz at Corning Glass Works produced the first fiber with losses under 20 dB/km. In doing so, they proved Kao and Hockham to be correct, but too conservative in their goals for fiber-loss levels. By 1972, losses in fibers were reduced to 4 dB/km in laboratory samples. Today, losses in the best fibers are under 0.2 dB/km, two orders of magnitude better than the levels recommended by Kao and Hockham. To give perspective on the importance of low loss, consider this: The 20 dB loss first achieved by Maurer, Keck, and Schultz means that only 1% of the optical energy injected into the fiber is left after a 1-km transmission distance. By 1982, fiber's performance improved dramatically enough that light could be transmitted over 125 km and still retain 1% of the optical energy.

The combination of laser and optical fiber sparked the imaginations of communications engineers worldwide. Here was a source of great potential and a transmission medium of uncommon promise. Similar advances in related technologies, such as detectors, transmission technology, and communication theory led to investigations of fiber-optic systems during the mid- and late 1970s.

The Navy installed a fiber-optic telephone link aboard the USS *Little Rock* in 1973. The Air Force replaced the copper wiring harness of an A-7 aircraft with an optical link in its Airborne Light Optical Fiber Technology (ALOFT) program. The copper harness contained 302 cables with a total length of 1250 m and a weight of 40 kg. The optical replacement has 12 fibers, a total length of 76 m, and a weight of 17 kg. The military was also responsible for one of the earliest operational fiber-optic links in 1977, a 2-km, 20 Mbps system connecting a satellite earth station and a computer center.

In 1977, both AT&T and GTE installed fiber-optic telephone systems carrying commercial telephone traffic. Encouragingly, the systems exceeded the rigid performance standards for reliability imposed by the telephone companies on their equipment and operations. As a result of successful early trials, optical systems were installed by telecommunications companies throughout the late 1970s and early 1980s. In 1980, AT&T announced plans for an ambitious project to construct a fiber-optic system along the Northeast Corridor from Boston to Richmond. The importance of this project lay in the arrival and acceptance of fiber as a mainstream technology for normal, high-capacity systems.

Technology advanced during the same period on every front as the industry matured and gained experience. As late as 1982, for example, advanced single-mode fibers (which will be described in Chapter 4) were felt to be so difficult to use that they would find only specialized applications for many years. MCI installed the first large-scale single-mode system in 1983, and by 1985, major long-distance carriers like AT&T and MCI announced that single-mode fibers were the fiber of choice for nearly all telecommunication uses.

Although the computer, information network, and industrial control industries did not embrace fiber optics as quickly as did the telecommunications industry, they nonetheless carefully investigated the technology, experimented with it, and closely observed the experiences of others. The needs of these other industries often required fiber-optic solutions different from those used in telecommunications. For example, the cost-effectiveness of fiber in a telephony application and a computer-network application are different. Still, fiber optics arrived in these other industries because it offered benefits that could not be easily obtained with copper cable.

In 1990, for example, IBM, the world's largest computer maker, announced a new mainframe computer. The channel controller, named ESCON, which links the computer to such peripherals as disk and tape drives, uses optical fibers as the cabling medium. This was the first time fiber was offered as *standard* equipment. The benefit of this fiber-based ESCON controller is that it allows faster transfer of information over longer distances. The copper-based controller it replaced had an average transfer rate of 4.5 Mbyte/s over a maximum distance of 400 feet. The ESCON system offered a transfer rate of 10 Mbyte/s over distances of several miles.

Another initiative from the late 1980s was the Fiber Distributed Data Interface (FDDI). FDDI was the first local area network designed specifically for fiber optics. It offered a data rate of 100 Mbps. At the time, the prevailing networks ran much slower: 10 Mbps for Ethernet and 16 Mbps for Token Ring. Today, FDDI runs on both fiber and copper, and Gigabit Ethernet operates at 1000 Mbps. Other systems run at even higher speeds.

Today, fiber optics is a mainstream technology still growing rapidly. The original fiber-optic telephone system installed by AT&T in Chicago in 1977 operated at 44.7 Mbps. By today's standards, that's slow. One basic rate for a fiber-optic telephone system is 2.5 Gbps, over five times faster. What's more, telephone companies are combining several 2.5-Gbps light streams over a single fiber to achieve a higher rate. In 1998, up to 16 channels were routinely transmitted to yield a data transmission rate of 40 Gbps, with even larger systems becoming available.

THE INFORMATION AGE

Our era is sometimes called the information age, which also suggests that ours is the electronics age. The information age uses electronics to perform the four basic functions of information processing:

1. Gathering information
2. Storing information

3. Manipulating and analyzing information
4. Moving information from one place to another

The tools of the age are fairly new: the computer, electronic offices, a complex telephone network, information networks and databases, cable television, the Internet, and so forth. The evidence surrounds you. The information services industry has been growing at a rate of 15% a year, with no signs of slowing down.

The United States—and much of the world—is becoming a huge network tied together by electronics. As easily as you can telephone from New York to San Francisco, you can transmit computer data. Engineers use computers to design new computers. Personal computers are now commonplace in the home. The Internet—a giant network of networks—has become a significant force in the 1990s.

The following facts give some perspective to the importance of electronics in modern life.

- The personal computer did not exist in 1976; today it is a standard piece of office equipment, as common as a telephone or fax machine. The biggest growth area for computers is in the home.
- More than 2,500 computer databases are available from which you can obtain information with a personal computer and public telephone network. In addition, the Internet has become an important source of information for researching everything from fiber-optic technology to new car purchases.
- When the previous edition of this book was written, you could use a computer and modem to communicate over the telephone network. In those days, high-speed communication was 14.4 kbps. Not only have modem speeds increased to 56 kbps, a number of other options provide access speeds of several megabits per second.
- In March 1997, the Internet got its one-millionth domain name. A domain name is the name of an Internet site (such as www.ibm.com). While the Internet traces its roots to the 1960s as a means for academic and government researchers to communicate, it experienced explosive growth in the early 1990s with the invention of the World Wide Web. The Web, with its graphic interface, makes it easier (and more fun) to use the Internet. Today the Internet is a growing means of conducting research, doing business, entertaining yourself, and communicating with others around the world.
- It is estimated that by 2000, Internet traffic on the telephone system will exceed the traffic from voice calls. By 2006, Internet traffic will require three times the bandwidth of that required for voice traffic.
- By 2000, there will be an estimated 6.6 trillion e-mail messages sent annually by over 80 million users.
- The first fiber-optic telephone system, installed in 1977, operated at 44.7 Mbps and was capable of carrying 672 voice calls simultaneously over a single

fiber. Today, telephone systems over fiber operate 400 times faster, at 20 Gbps and several hundred thousand voice calls.

Information may be a call to a friend, the design of a new widget, or stock quotes found on the Internet. Because of the growing amount of information, the means of moving it from one place to another is important. Telephone companies use digital means to send both your voice and computer data. Surprisingly, your voice becomes more like computer data: it is converted to digital pulses—numbers—just like computer data. By making this conversion, the telephone company transmits your voice with greater fidelity and reliability. Digital telephony has replaced analog methods almost completely in the United States. The only transmission still analog is between the home and the telephone switching office. Nearly all switching offices digitize your voice before retransmitting it.

Fiber optics is exceptionally suited to the requirements of digital telecommunications. Coupled with the need to move information is the need to move it faster, more efficiently, more reliably, and less expensively. Fiber optics fills these needs.

The Wiring of America

The oldest and most obvious example of the wiring of the United States is the telephone network. At one time, the network carried voices almost exclusively—it now carries voice, computer data, electronic mail, and video pictures. As mentioned above, Internet traffic is expected to exceed voice traffic by the year 2000.

Cable television is another example. It offers fifty or more channels, with hundreds of channels coming soon. The cable television infrastructure can also be used for telephone services and Internet access at speeds of up to 30 Mbps.

Home banking and financial services are also made possible by the wiring of America. From a personal computer or terminal in your home, you can check account balances, transfer funds between accounts, pay bills, and buy stocks and bonds.

The office is also wired. Computers have replaced the secretary's typewriter, the accountant's balance sheet, the engineer's drafting table, and the artist's pens and airbrush. Furthermore, a need exists to allow computers to communicate with each other and to share resources such as printers, fax machines, files, and access to the Internet. Local area networks (LANs) fill this need by allowing equipment to be tied together electronically. The local network can, in turn, be connected to a wide area network (WAN). Businesses use WANs to tie remote offices to headquarters or to connect a LAN to the Internet.

Telecommunications and the Computer

Until the 1980s, a clear distinction existed between what was part of the telephone system and what was a computer system. As the world's largest network, the public telephone network once carried voice traffic almost exclusively. Computer data represented only a small portion of what was carried. Today, not only are voices digitized into data, the Internet and other computer data make up a significant portion of the traffic and are expected to exceed voice traffic by the end of this decade. Internet access can be offered over cable TV or direct satellite broadcasting. Cable TV providers are looking at becoming suppliers of telephone service, while telephone companies would like to provide enhanced services including television, Internet, and movies. Even the power utilities are considering ways to offer services. In the future, you will probably buy services from a provider whose title does not fit traditional categories.

One thing that has made this possible is digital transmission of information, which we will discuss further in the next chapter. Traditional analog TV is scheduled to be obsolete by 2006, with digital TV replacing it. The list of digitally transmitted information is growing—from music CDs to the Internet and television.

Bandwidth makes this digital revolution possible. For all the advantages of digital transmission, tremendous capacity is required. Fiber optics is an enabling technology providing the bandwidth to make this digital information revolution practical. Compared to copper cables, optical-fiber cables can carry more information further. That is, they provide greater bandwidth over longer distances. In essence, that is the reason for fiber optics. The technology has additional advantages, but high-bandwidth, long-distance transmission is what first intrigued researchers and what led to its widespread use today.

What does this bandwidth/distance advantage mean in practical terms? In the 1970s, a standard telecommunication cable had the capacity to carry 672 simultaneous voices about 2 km before the signal had to be amplified. Today, a single fiber can carry over 130,000 simultaneous voices over 60 km. Cable television providers, too, are replacing coaxial trunks with fiber. A single fiber can not only replace the cable, but eliminate the need for 35 to 45 amplifiers along the way—and deliver more channels.

In telecommunications, CATV, computer networking, and a host of other applications, the optical fiber and laser (and the closely related light-emitting diode) have proven a powerful combination in realizing the information age.

SUMMARY

- The four parts of a basic fiber-optic link are the transmitter, cable, receiver, and connectors.
- Using electronics to gather, store, manipulate, and move information has resulted in the networking of this country.
- The practical use of fiber optics for communications began in the mid- and late 1970s with field trials. Today fiber is an accepted technology.
- The optical fiber is an ideal transmission medium for information.
- Information is increasingly transmitted in digital form.

Review Questions

1. Name the three main parts of a fiber–optic system.
2. While optical fibers were invented in the early part of this century, they did not find use as a communications medium for one main reason. What is the reason?
3. Which of the following is the main purpose of the fiber?
   - **A.** To channel the light into a circular shape
   - **B.** To guide the light
   - **C.** To change the wavelength of the light from visible to infrared
   - **D.** To amplify the output of a laser
4. The first fiber-optic telephone system operated at what data rate?
5. In the public telephone network, which of the following represent the biggest growth in traffic?
   - **A.** Long-distance telephone calls
   - **B.** Faxes
   - **C.** The Internet
   - **D.** Video
6. Name the four functions of information processing.
7. Of the four functions of information processing, which one makes use of fiber-optic technology?
8. What was the main obstacle to overcome in achieving a low-loss optical fiber?
9. In the 1970s, the greatest interest in applying fiber optics came from which of the following industries?
   - **A.** Military/government
   - **B.** Computer industry
   - **C.** Avionics industry
   - **D.** Telephone industry
   - **E.** Automotive industry
   - **F.** Medical industry
   - **G.** Power industry
10. If you have Internet access at home or work, at what rate do you connect? Is this a good use of fiber? Why?

chapter two

INFORMATION TRANSMISSION

The last chapter discussed the importance of electronic information to our modern world. The purpose of this chapter is to introduce important aspects of signals and their transmission. An understanding of the underlying principles that allow modern electronic communication is fundamental to an understanding and appreciation of fiber optics. The ideas presented here are fundamental not only to fiber optics but to all electronic communication. We do not propose to offer an in-depth survey of communication theory and its application to electronics. The purpose of this chapter is to introduce terms and principles without which any discussion of fiber optics would be stymied.

COMMUNICATION

Communication is the process of establishing a link between two points and passing information between them. The information is transmitted in the form of a signal. In electronics, a signal can be anything from the pulses running through a digital computer to the modulated radio waves of an FM radio broadcast. Such passing of information involves three activities: encoding, transmission, and decoding.

Encoding is the process of placing the information on a carrier. The vibration of your vocal chords places the code of your voice on air. The air is the carrier, changed to carry information by your vocal chords. Until it is changed in some way, the carrier carries no information. It is not yet a signal. A steadily oscillating electronic frequency can be transmitted from one point to another, but it also carries no intelligence unless information is encoded on it in some way. It is merely the carrier onto which the information is placed. Conveying information, then, is the act of modifying the carrier. This modification is called *modulation*.

Figure 2-1 shows the creation of a signal by impressing information on a carrier. A high-frequency carrier, which in itself carries no information, has impressed on it a lower-frequency signal. The shape of the carrier is now modulated by the information. Although the simple example in the figure conveys very little information, the concept can be extended to convey a great deal of information. A Morse-code system can be based on the example shown. On the unmodulated carrier can be impressed a low frequency for one of two durations corresponding to the dots and dashes of the code.

Once information has been encoded by modulating a carrier, it is *transmitted*. Transmission can be over air, on copper cables, through space to a satellite and back, or through optical fibers.

At the other end of the transmission, the receiver separates the information from the carrier in the *decoding* process. A person's ear separates the vibrations of the air and turns them into nerve signals. A radio receiver separates the information from the carrier. For the encoded signal in Figure 2-1, the receiver would strip away and discard the high-frequency carrier while keeping the low-frequency signal for further processing. In fiber optics, light is the carrier on which information is impressed.

Assume you are writing a letter on a personal computer to a friend. The letters you press at the keyboard are encoded into the digital pulses the computer requires. Once finished, you wish to send the letter to your friend's computer. You will need a *modem,* a device that re-encodes the digital pulses into audio pulses easily sent over the telephone lines to the central telephone office.

The central office receives your transmission. It reencodes your signal into digital pulses and transmits it with many other signals over an optical fiber to another central office. This office decodes the digital pulses to audio pulses. It sends this information to your friend, whose modem further decodes the signal into computer pulses. Your friend can read your letter by displaying it on the computer screen or by printing it out.

FIGURE 2-1
A signal

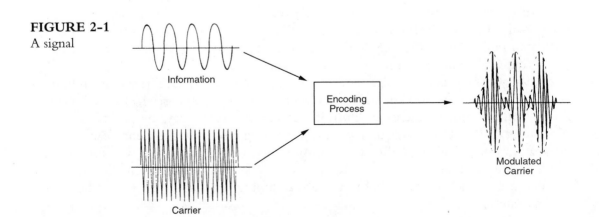

Information

Encoding
Process

Modulated
Carrier

Carrier

Many ways exist to modulate the carrier. Figure 2-2 shows three common ways.

1. *Amplitude modulation (AM).* Amplitude modulation is used in AM radio. Here the amplitude of the carrier wave is varied to correspond to the amplitude of the information.
2. *Frequency modulation (FM).* Frequency modulation changes the frequency of the carrier to correspond to differences in the signal amplitude. The signal changes the frequency of the carrier, rather than its amplitude. FM radio uses this technique.
3. *Pulse-coded modulation (PCM).* Pulse-coded modulation converts an analog signal, such as your voice, into digital pulses. Your voice can be represented as a series of numbers, the value of each number corresponding to the amplitude of your voice at a given instance. PCM is the main way that voices are sent over a fiber-optic telephone system. We will look more closely at it in a moment.

ANALOG AND DIGITAL

We live in an analog world. *Analog* implies continuous variation, like the sweep of the second hand around the face of an electric watch. Your voice is analog in that the vibrations that form sound can vary to any point within your range of vocal frequencies. Sound

FIGURE 2-2
Examples of
modulation
(Illustration
courtesy of AMP
Incorporated)

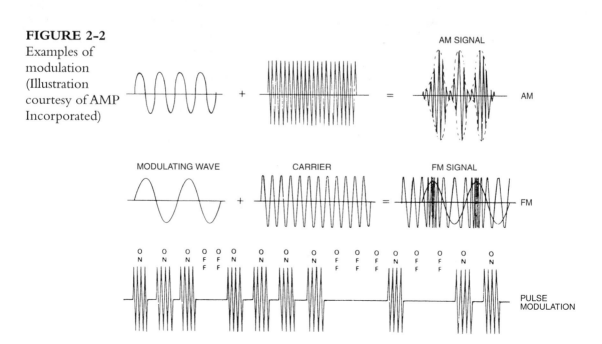

is analog, in that it is continuously varying within a given range. Electronic circuits, such as transmitters and receivers, use analog circuits to deal with these continuous variations. Indeed, before the advent of computers, most electronic engineering dealt with analog circuits.

Consider a light operated by a dimmer. Turning the dimmer to adjust the light's brightness is an analog operation. The brightness varies continuously. There are no discrete levels, so that you can easily turn the light more or less bright.

In contrast, *digital* implies numbers—distinct units like the display of a digital watch. In a digital system, all information ultimately exists in numerical values of digital pulses.

In contrast to the dimmer light, a three-way lamp is digital. Each setting brings a specific level of brightness. *No levels exist in between.* Figure 2-3 shows analog and digital signals.

Important to electronic communication is that analog information can be converted to digital information, and digital information can be turned into analog information. You may be familiar with digital stereo systems. In these systems, music is encoded into digital form as a series of numbers that represent the analog variations in the music. The electronics in the playback system must reconvert the digital code into analog music.

Digital Basics: Bits and Bytes

The basis of any digital system is the *bit* (short for *bi*nary dig*it*). The bit is the fundamental unit of digital information, and it has only one of two values: 1 or 0. Many ways exist to represent a bit. In electronics, the presence or absence of a voltage level is most common: One voltage level means a *1;* a second level means a *0.* Unfortunately, the single bit 1 or 0 can represent only a single state, such as *on* or *off.* For example, a lamp can be represented by 0 when turned off and by 1 when turned on:

Off   = 0
On   = 1

A single bit of information, then, is of very limited usefulness. But we can describe the state of the three-way lamp with 2 bits:

Off   = 00
On   = 01
Brighter = 10
Brightest = 11

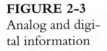

FIGURE 2-3
Analog and digital information

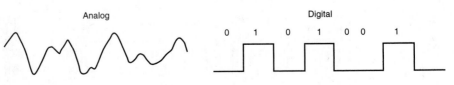

Two bits allow us to communicate more information than we can with a single bit. In our lamp example, 2 bits allowed us to distinguish between four distinct states. The more bits we use in a unit, the more potential information we can express. A digital computer typically works with units of 8 bits (or multiples of 8, such as 16 and 32).

An 8-bit group is called a *byte*. A single byte allows all the letters, numbers, and other characters of a typewriter or computer keyboard to be encoded by a number, with room left over. Eight bits permits 256 different meanings to be given to a pattern of 1s and 0s. The number of different combinations or meanings for any given length of bits *n* is equal to 2 to the *n*th power, 2^n. For example, 16 bits yields 65,536 combinations, since $2^{16} = 65,536$. Each time a bit is added, the number of possible different values doubles.

A pulse train is often shown in its ideal form, as the string in Figure 2-4. The pulses go from one state to another instantaneously. Such diagrams show ideal pulses and the main essentials of the pulse train, so engineers and technicians can compare trains.

A pulse train represents the 1s and 0s of digital information. The train can depict high- and low-voltage levels or the presence and absence of a voltage. In electronics, a digital 1 can be a voltage pulse or a higher voltage level. A digital 0 can be the absence of the pulse or a lower voltage level. Thus we can speak of a 1 with terms such as *on* or *high* and of 0 with terms such as *off* or *low*.

A real pulse does not occur instantaneously as it is shown doing in Figure 2-4. An electronic circuit has a finite response time—that is, it takes time for a voltage or optical pulse to turn on and off or switch between high and low levels. The pulse must also stay on for a brief time. Still, in a computer system, turning a pulse on and off can require only millionths or billionths of a second, to allow thousands or millions of pulses per second.

Engineers dealing with digital systems must consider the shape of the pulse. Figure 2-5 defines the parts of the pulse.

- *Amplitude* is the height of the pulse. It defines the level of energy in the pulse. The energy can be voltage in a digital system or optical power in a fiber-optic system. Notice that different types of energy are used for different types of systems.

- *Rise time* is the time required for the pulse to turn on—to go from 10% to 90% of its maximum amplitude.

FIGURE 2-4
Ideal pulse train
(Illustration
courtesy of AMP
Incorporated)

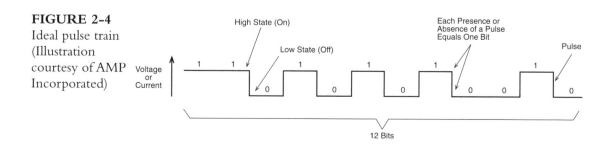

FIGURE 2-5

Parts of a pulse

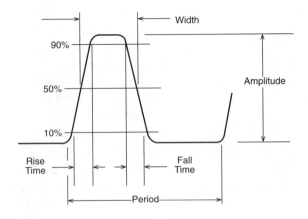

- *Fall time,* the opposite of rise time, is the time required to turn off, measured from 90% to 10% of amplitude. Rise time and fall time may or may not be equal.
- *Pulse width* is the width, expressed in time, of the pulse at 50% of the amplitude.
- *Bit period* is the time given for the pulse. Most digital systems are clocked or timed. Pulses must occur in the time allotted by the system for a bit period. For example, suppose a string of five 0s occurs. How do we know it is not four 0s or six 0s? We can detect no changes in the pulse train. We know because nothing has occurred in 5 bit periods of the clock.

A clock is an unchanging pulse train that provides timing by defining bit periods. Each bit period can be defined as one or more clock timing periods. The clock is loosely comparable to a musician's metronome: Both provide the basic timing beat against which operation (pulses or musical notes) is compared.

Rise time is especially important in electronics and fiber optics. Rise time limits the speed of the system. The speed at which pulses can be turned on and off will determine the fastest rate at which pulses can occur. The easiest way to increase the speed of a system is to decrease pulse rise and fall times, thereby turning pulses on and off faster. Thus more pulses can occur in a given time. Even when pulse amplitude and pulse width remain the same, faster rise times bring faster operating speeds. In other cases, we wish to maintain the pulse width as a certain proportion of pulse time. Faster rise and fall times allow shorter pulse widths, which increase operating speed even more. Conversely, slowing the rise time slows the operating speed.

As we will see in Chapter 10, the bits of 1s and 0s can be formed in other ways than simply turning a voltage on and off. Different formats for coding 1s and 0s offer various advantages and disadvantages in transmitting information. Here we have showed a 1 as a high voltage and a 0 as a low voltage occurring in a given bit period. Other codes use both high and low voltages within a bit period to encode a 1 or 0. Still, the essential thing here is that digital systems use 1s and 0s, high and low states, to encode information.

Why Digital?

Computers use digital signals because it is their nature: They are digital by design. Although a computer does not have to use digital techniques and binary numbers, it does so because they are the most convenient way to construct a computer that operates electronically. Transistors and integrated circuits made up of thousands of transistors operate as very fast on/off switches.

Telephone companies use digital techniques and turn your voice into a digital signal before transmitting it. What is the advantage? A signal transmitted any distance becomes distorted. Even if your voice is faithfully reproduced electronically for transmission as an analog signal, it still is distorted by the time it reaches the receiver. Unfortunately, the receiver cannot correct the distortion, because it has no way of knowing what the original signal looked like.

For digital pulses, the situation is reversed. A digital pulse has a defined shape. The receiver knows what the original pulse looked like. A receiver only needs to know information about the pulses: How many pulses? When were they sent? and so on. The receiver can be designed to rebuild distorted pulses into their original shape to faithfully reproduce the original signal.

INFORMATION-CARRYING CAPACITY

Any transmission path carrying signals has limits to the amount of information it can carry. The amount of information that the path can carry is its *information-carrying capacity*. Several ways exist to describe this capacity. In telephony, capacity is expressed in voice channels. A voice channel is the bandwidth or range of frequencies required to carry a single voice. Since the upper end of the human voice is about 4 kHz, a single voice channel must have a bandwidth of 4 kHz. In the early days of telephones, each wire carried only a single voice. Today, a single telephone line carries hundreds or thousands of voices simultaneously. It thus has a capacity of thousands of voice channels.

Bandwidth is proportional to the highest rate at which information can pass through it. If an optical fiber, for example, has a bandwidth of 400 MHz, it can carry frequencies up to this limit.

In digital systems, capacity is given in bits per second (bps) or *baud*. For a telephone system, a single digital voice channel requires 64,000 bps. A digital system, therefore, requires more bandwidth than a comparable analog system. An analog telephone system requires 4 kHz per voice; a digital system requires over 16 times more—64 kHz. Simple digital telephone systems, which carry 672 voices one way over a single line, have an operating speed of 44.7 megabits per second (Mbps).

The methods that allow such systems to be built are PCM and multiplexing.

PCM AND MULTIPLEXING

The technique used to turn your analog voice into a digital signal is *pulse-code modulation* (PCM). Time-division multiplexing is the means that allows several voice channels to be transmitted on a single line.

Communication theory states that an analog signal, such as your voice, can be digitally encoded and decoded if it is sampled at twice the rate of its highest frequency. The high end of the speaking voice in telephony is 4000 Hz, which means that a PCM system must sample the voice 8000 times a second. *Sampling* means that the system looks at the amplitude of the vocal frequencies. Each sample is converted into an 8-bit number. Eight bits allow 256 different amplitudes to be encoded. Since each sample requires 8 bits and the sampling occurs 8000 times a second, the bandwidth required for a single voice channel is 64,000 bps (8000 samples × 8 bits/sample = 64,000 bits). At the receiver end, the same rules used to encode the sample are used to decode the sample and to reconstruct the analog voice signal. Figure 2-6 shows the idea of PCM. There are many ways to design and use PCM systems, but the basic idea is as we have described.

Sending 64,000 bps does not make efficient use of the capacity of a typical transmission path. To make full use of the capacity, telephone companies send several voice channels over the path. The appearance is that all channels are transmitted simultaneously, although they really are not. Part of one voice is first sent, and then part of the second, part of the third, and so on, in an interleaving pattern of channels. The device that permits this combination of different signals onto a single line is a *multiplexer*. The speed of the system easily allows all signals to be sent in turn. A *demultiplexer* at the receiver performs the opposite operation and separates the signals.

FIGURE 2-6
PCM
(Illustration
courtesy of AMP
Incorporated)

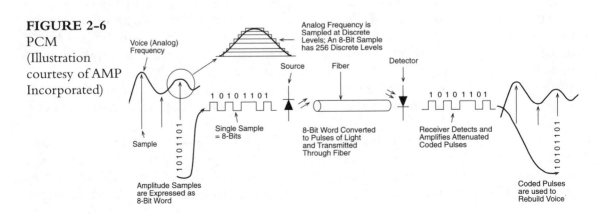

Time-division multiplexing (TDM) is the name given to the method of multiplexing that assigns parts of each voice channel to specific time slots. Other forms of multiplexing, such as *frequency-division multiplexing* (FDM) and *wavelength-division multiplexing* (WDM), exist. FDM assigns different information channels to different carrier frequencies. Cable television uses FDM. WDM, which is unique to optical communications, is discussed in Chapters 12 and 16.

Fiber optics is important to such telephone systems because its information-carrying capacity exceeds wire-based systems.

THE DECIBEL

The *decibel* is an important unit that you will use continually in fiber optics as well as in electronics. It is used to express gain or loss in a system or component. A transistor, for example, can amplify a signal, making it stronger by increasing its voltage, current, or power. This increase is called *gain.* Similarly, loss is a decrease in voltage, current, or power. The basic equations for the decibel are

$$dB = 20 \log_{10} \left(\frac{V_1}{V_2} \right)$$

$$dB = 20 \log_{10} \left(\frac{I_1}{I_2} \right)$$

$$dB = 10 \log_{10} \left(\frac{P_1}{P_2} \right)$$

where V is voltage, I is current, and P is power. The decibel, then, is the ratio of two voltages, currents, or powers. Notice that voltage and current are 20 times the logarithmic ratio, and power is 10 times the ratio.

A basic use of the decibel is to compare the power entering a system, circuit, or component to the power leaving it. The decibel value shows how the device affects the circuit. A transistor, for example, usually increases the power. Other components may decrease the power. P_1 is the power out, and P_2 is the power in.

$$dB = 10 \log_{10} \left(\frac{P_{out}}{P_{in}} \right)$$

$$P_{in} \rightarrow \boxed{circuit} \rightarrow P_{out}$$

Another use of the decibel is to describe the effect of placing a component in a system. Sometimes, for example, it is necessary to split a cable and apply connectors so that the two cable halves can be connected and disconnected. Insertion of these connectors causes some loss, which is expressed in decibels. In other cases, devices called *attenuators* are purposely placed in a circuit to provide loss.

In fiber optics, we deal mostly with loss and optical power. (The electronic circuits in the transmitter and receiver may use voltage and current and provide gain.) The source emits optical power. As light travels through the fiber to the receiver, it loses power. This power loss is expressed in decibels. For example, if the source emits 1000 microwatts (μW) of power and the detector receives 20 μW, the loss through the system is about 17 dB:

$$
\begin{aligned}
\text{Loss} &= 10 \log_{10}\left(\frac{P_r}{P_{tr}}\right) \\
&= 10 \log_{10}\left(\frac{20}{1000}\right) \\
&= -16.989 \text{ dB}
\end{aligned}
$$

where P_{tr} is the power transmitted from the source and P_r is the power received by the receiver.

Figure 2-7 shows the amount of power remaining for different decibel values. A 10-dB loss (−10 dB) represents a loss of 90% of the power; only 10% remains. An additional loss of 10 dB increases the loss by an order of magnitude. A useful figure to remember is 3 dB, which represents a loss of one half of the power.

Fiber-optic links easily tolerate losses of 30 dB, meaning that 99.9% of the power from the source is lost before it reaches the detector. If the source emits 1000 μW of power, only 1 μW reaches the detector. Of concern in fiber optics are the proper balance of the source power, the losses in the system, and the sensitivity of the detector to weak light signals.

FIGURE 2-7
Loss and the decibel

LOSS (dB)	POWER REMAINING (%)	LOSS (dB)	POWER REMAINING (%)
0.1	97.7	4	39.8
0.2	95.5	5	31.6
0.3	93.3	6	25.1
0.4	91.2	7	19.9
0.5	89.1	8	15.8
0.6	87.1	9	12.6
0.7	85.1	10	10.0
0.8	83.2	20	1.0
0.9	81.1	30	0.1
1	79.4	40	0.01
2	63.1	50	0.001
3	50.1	60	0.001

Remember that a decibel expressing loss is a negative unit. In fiber optics, it is common practice to omit the negative sign and speak of a loss of 6 dB, say. This loss is actually −6 dB. If you solved the equation, the result would be −6 dB. But in talking, and even in data sheets, the negative sign is omitted, with little confusion being caused. But if you have occasion to use a loss figure in an equation, do not forget to make the number negative! (The confusion arises because some equations have been adjusted to accept a loss figure as a positive number.)

Sometimes the ratio for calculating loss or gain uses a constant P_2. In fiber optics, this value is usually 1 milliwatt (mW).

$$\text{dBm} = 10 \log_{10} \left(\frac{P}{1 \ mW} \right)$$

dBm means "decibels referenced to a milliwatt." In this case, the negative sign is almost always used. A value of −10 dBm means that P is 10 dB less than 1 mW, or 100 μW. Similarly, −3 dBm is 500 μW. Communication engineers and technicians use dBm units extensively.

Figure 2-8 relates various dBm values to milliwatt and microwatt levels. Notice that a small range of the dBm values allows a great range of power values to be expressed.

Another unit sometimes used is dBμ—decibel referenced to 1 μW. This is the same as dBm, except the reference value is 1 μW rather than 1 mW.

$$\text{dBμ} = 10 \log_{10} \left(\frac{P}{1 \ \mu W} \right)$$

FIGURE 2-8
Power-to-dBm
conversion

10 mW	=	+10 dBm
5 mW	=	+7 dBm
1 mW	=	0 dBm
500 μW	=	−3 dBm
100 μW	=	−10 dBm
50 μW	=	−13 dBm
10 μW	=	−20 dBm
5 μW	=	−23 dBm
1 μW	=	−30 dBm
100 nW	=	−40 dBm
10 nW	=	−50 dBm
1 nW	=	−60 dBm
100 pW	=	−70 dBm
10 pW	=	−80 dBm

SUMMARY

- Communication consists of encoding, transmitting, and decoding information.
- Digital systems work with bits representing 1 and 0.
- Information-carrying capacity expresses the amount of information that can be transmitted over a communication path.
- PCM turns an analog signal into a digital signal.
- Multiplexing allows many signals to be sent over a single communications path.
- The decibel is the primary unit for indicating gain or loss in a system.

 Review Questions

1. Name the three activities involved in communication. Give an example of each.
2. Define modulation.
3. Sketch an analog signal. Sketch a digital signal.
4. Is the information 1101101 analog or digital?
5. Sketch a pulse. Label its amplitude, rise time, fall time, and pulse width.
6. What is the capacity of an analog voice channel? Of a digital voice channel?
7. For a PCM system as described in this chapter, how many bits per second are required to transmit five voice channels? 50 voice channels?
8. Name the technique that allows five voice channels to be transmitted simultaneously over a single optical fiber.
9. Which requires greater bandwidth, an analog voice channel or a digital voice channel?
10. How many milliwatts are in a signal having a power of 0 dBm?

chapter three

FIBER OPTICS AS A COMMUNICATIONS MEDIUM

its advantages

In its simplest terms, fiber optics is a communications medium linking two electronic circuits. The fiber-optic link may be between a computer and its peripherals, between two telephone switching offices, or between a machine and its controller in an automated factory. Obvious questions concerning fiber optics are these: Why go to all the trouble of converting the signal to light and back? Why not just use wire? The answers lie in the following advantages of fiber optics:

- Wide bandwidth
- Low loss
- Electromagnetic immunity
- Light weight
- Small size
- Safety
- Security

The importance of each advantage depends on the application. In some cases, wide bandwidth and low loss are overriding factors. In others, safety and security are the factors that lead to the use of fiber optics. The following pages discuss each advantage in detail.

WIDE BANDWIDTH

The last chapters showed the increasing concern with sending more and more information electronically. Potential information-carrying capacity increases with the bandwidth of the transmission medium and with the frequency of the carrier. From the earliest days of radio, useful transmission frequencies have pushed upward five orders of magnitude, from about 100 kHz to about 10 GHz. The frequencies of light are several orders of magnitude above the highest radio waves. The invention of the laser, which uses light as a carrier, in a single step increased the potential range four more orders of magnitude—to 100,000 GHz (or 100 terahertz [THz]). Optical fibers have a potential useful range to about 1 THz, although this range is far from being exploited today. Still, the practical bandwidth of optical fibers exceeds that of copper cables. Furthermore, the information-carrying possibilities of fiber optics have only begun to be exploited, whereas the same potentials of copper cable are pushing their limits.

As mentioned earlier, telephone companies increasingly use digital transmission. The higher bandwidth of optics allows a higher bit rate and, consequently, more voice channels per cable. For compatibility to exist among all telephone carriers, the rates at which different transmission lines carry information are generally fixed by a system known as the North American Digital Telephone Hierarchy.

Figure 3-1 shows the hierarchy for coaxial cables and optical fibers. The coaxial system has been long established. The fiber-optic capacities shown are for SONET, or synchronous optical network. Chapter 16 describes SONET in greater detail.

One consequence of fiber's high bandwidth is that it permits transmission of channels requiring much greater bandwidth than a voice channel. Television and teleconferencing, for example, require a channel capacity 14 to 100 times that of a digitally encoded voice. The bandwidth of a fiber allows these signals to be multiplexed through the fiber, permitting voice, data, and video to be transmitted simultaneously. The demand for these services means that fibers will move from being only long-distance carriers to being carriers right to the home and business.

To give perspective to the incredible capacity that fibers are moving toward, a 10-Gbps signal has the ability to transmit any of the following *per second:*

- 1000 books
- 130,000 voice channels
- 16 high-definition television (HDTV) channels or 100 HDTV channels using compression techniques. (An HDTV channel requires a much higher bandwidth than today's standard television.)

Similar advances will not occur for coaxial systems: Optical systems will become the favored solution for long-distance, high-data-rate applications.

Another consequence of the high bandwidth of fiber optics is that it permits transmission of channels requiring much greater bandwidth than a voice. Television and teleconferencing require a channel capacity 14 to 100 times that of a digitally encoded voice. The bandwidth of the fiber allows these signals also to be multiplexed through the fiber. Fibers allow telephone companies to simultaneously transmit voice, data, and video, which is important to the information age.

FIGURE 3-1
Digital telephone transmission rates

MEDIUM	BIT RATE DESIGNATION	(Mbps)	VOICE CHANNELS	REPEATER SPACING (km)
Coaxial Cable	DS-0	0.064	1	1-2
	DS-1	1.544	24	
	DS-1C	3.152	48	
	DS-2	6.312	96	
	DS-3	44.736	672	
Fiber (SONET)	OC-1	51.84	672	40+ (Laser)
	OC-3	155.52	2016	2 (LED)
	OC-12	622.08	8064	
	OC-48	2488.32	32,256	
	OC-96	4976.64	64,512	
	OC-192	9953.28	129,024	

LOW LOSS

Bandwidth is an effective indication of the rate at which information can be sent. Loss indicates how far the information can be sent. As a signal travels along a transmission path, be it copper or fiber, the signal loses strength. This loss of strength is called *attenuation*. In a copper cable, attenuation increases with modulation frequency: The higher the frequency of the information signal, the greater the loss. In an optical fiber, attenuation is flat; loss is the same at any signaling frequency up until a very high frequency. Thus, the problem of loss becomes greater in a copper cable as information-carrying capacity increases.

Figure 3-2, by showing the loss characteristics for fibers, twisted pairs, and coaxial cable, such as are used in telephone systems, demonstrates the usable ranges for these transmission media. Loss in coaxial cable and twisted-pair wires increases with frequency, whereas loss in the optical cable remains flat over a very wide range of frequencies. The loss at very high frequencies in the optical fiber does not result from additional attenuation of the light itself by the fiber. This attenuation remains the same. The loss is caused by loss of information, not by optical power. Information is contained in the variation of the optical power. At very high frequencies, distortion causes a reduction or loss of this information.

The point is that the effects of loss that must be accounted for in a system depend on the signal frequency. What is suitable for a system working at one speed may not work at another frequency. This need to consider different signaling speeds complicates designs.

FIGURE 3-2

Attenuation versus frequency (Courtesy of Siecor Corporation)

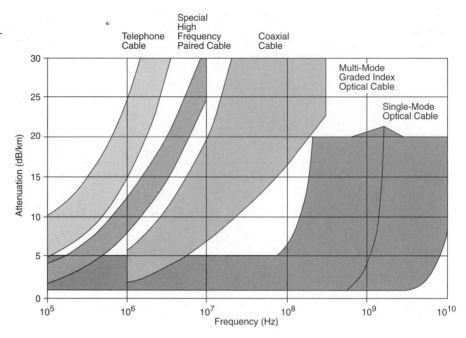

High-frequency design, for example, is more difficult than low-frequency design. One cannot simply increase operating speeds without taking into account the effects of speed on performance. In a fiber-optic system, the loss is the same at all operating speeds within the fiber's specified bandwidth.

Severe attenuation requires repeaters at intermediate points in the transmission path. For copper cables, repeater spacings, in general, decrease as operating speeds increase. For fibers, the opposite is true: Repeater spacings increase along with operating speeds, because at high data rates very efficient, low-loss fibers are used.

The first transatlantic fiber-optic telephone link, installed by AT&T in 1988, carried 37,800 simultaneous voice conversations each direction on a pair of fibers. Repeater spacings are 35 km. In contrast, the best transatlantic coaxial system handles 4200 conversations and has a repeater spacing of only 9.4 km. The possibility exists that fiber-optic systems will carry trillions of bits over 80 to 100 km without a repeater.

The combination of high bandwidth and low loss has made the telephone industry probably the heaviest user of fiber optics. It allows them not only to send more information over a transmission path but also to extend repeater spacings. Repeaters, after all, are electronics systems that are costly to build, install, and maintain. Fewer repeaters mean less costly systems.

ELECTROMAGNETIC IMMUNITY

Unlike copper cables, optical fibers do not radiate or pick up electromagnetic radiation. Any copper conductor acts like an antenna, either transmitting or receiving energy. One piece of electronic equipment can emit electromagnetic interference (EMI) that disrupts other equipment. Among reported problems resulting from EMI are the following:

- The military reported such a high concentration of electronic equipment in Vietnam that increases in the population of electronic devices beyond a crucial density made it impossible for the equipment to operate.
- An electronic cash register interfered with aeronautical transmissions at 113 MHz.
- Coin-operated video games interfered with police radio transmissions in the 42-MHz band.
- Some personal computers tested by the Federal Communications Commission (FCC) in 1979 emitted enough radiation to disrupt television reception several hundred feet away.
- Electrostatic discharges (ESD), discharged into computer terminals by operators, have garbled the computer memory, erased work in progress, and even destroyed circuits. (ESD is the shock you get when you walk across a carpet on a dry day and touch a doorknob. Such discharges easily contain 15 to 25 kV).
- An explosion caused by static electricity killed three workers at Cape Kennedy in the 1960s.
- A manufacturer of gasoline pumps found that CB radio transmissions zeroed remote pump readouts.
- Airport radar erased taxpayers' records in a computer memory bank.

EMI is a form of environmental pollution with consequences ranging from the merely irksome to the deadly serious. As the density of electronic equipment grows (and it will as part of the information age), the potential for EMI problems also increases. To combat such problems, the FCC in 1979, for example, issued stringent regulations limiting EMI from computing devices. Regulatory agencies in Europe have issued similar regulations.

Here is a simple way to demonstrate the effects of EMI if you have access to a personal computer or computer terminal. Place an AM radio near the computer while it is running. Tune through the dial. At some point, the radio should pick up the computer, so that you will be able to listen to the computer operate. Try running different programs to hear differences. What you hear is EMI.

Cables interconnecting equipment can be one of the main sources of EMI. They can also be one of the main receiving antennas carrying EMI into equipment. Cables act just like the radio antenna in the foregoing EMI demonstration.

Since fibers do not radiate or receive electromagnetic energy, they make an ideal transmission medium when EMI is a concern. Some factories use fiber optics because of its immunity. These applications do not require the high bandwidth or low loss of fibers. Equipment such as motors that switch on and off can be a source of EMI that disrupts signal lines controlling the equipment. Using fibers rather than copper cables eliminates the problems.

High-voltage lines can present another problem, since they also emit energy. Copper signal cables cannot be run next to such lines without special precautions, because energy from the high-voltage line couples onto the signal line. Fiber-optic lines can be run beside high-voltage lines with no detrimental effect, since no energy couples onto them from the high-voltage line.

One consequence of the fiber's electromagnetic immunity is that signals do not become distorted by EMI. Digital transmission requires that signals be transmitted without error. EMI can be a cause of error in electrical conductor transmission systems. A burst of EMI may appear as a pulse, where no pulse occurred in the original pulse stream. Fibers offer very high standards in error-free transmission.

LIGHT WEIGHT

A glass fiber weighs considerably less than a copper conductor. A fiber-optic cable with the same information-carrying capacity as a copper cable weighs less than the copper cable because the copper requires more lines than the fiber. For example, a typical single-conductor fiber-optic cable weighs 9 lb/1000 ft. A comparable coaxial cable weighs nine times as much—about 80 lb/1000 ft. Weight savings are important in such applications as aircraft and automobiles.

SMALL SIZE

A fiber-optic cable is smaller than its copper counterparts. In addition, a single fiber can often replace several copper conductors. Figure 3-3 shows a comparison of coaxial cable and a fiber-optic cable used in digital telephony. The copper cable is 4.5 in. in diameter and can carry as many as 40,300 two-way conversations over short distances. The fiber-optic cable, which contains 144 fibers in its 0.5-in.–diameter structure, has the capacity to carry 24,192 conversations on each fiber pair or nearly 1.75 million calls on all the fibers. The fiber-optic cable greatly exceeds the capacity of the coaxial cable even though it is almost 10 times smaller.

FIGURE 3-3
Size comparison: multiconductor coaxial cable and fiber-optic cable (Courtesy of AT&T/Lucent Technologies Bell Labs)

The small size of fiber-optic cables makes them attractive for applications where space is at a premium:

- Aircraft and submarines, where the use of every square inch is critical. Not only do the fiber-optic cables use less of the valuable space, they can be placed in areas where copper cables cannot be. In short, they make efficient use of the space.
- Underground telephone conduits, especially in cities. Here, not only are conduits often filled to capacity, but building new conduits to allow expanded services is very costly. Additional space may be unavailable. Fiber-optic cables replace the copper cables, often offering greater capacity in less space. A large copper cable that fills a conduit can be replaced by a smaller fiber cable, with room left for new cables in the future.
- Computer rooms, where the cables between equipment run under raised floors. These cables are often very rigid and difficult to install. Adding new cables is also difficult. Again, the small size and resulting flexibility of fiber eliminate such problems. Indeed, in some cases, so few fibers are needed that the need for a raised floor can also be eliminated.

SAFETY

A fiber is a dielectric—it does not carry electricity. It presents no spark or fire hazard, so it cannot cause explosions or fires as a faulty copper cable can. Furthermore, it does not attract lightning. The fiber-optic cable can be run through hazardous areas, where electrical codes or common sense precludes the use of wires. It is possible, for example, to run a fiber directly through a fuel tank.

SECURITY

One way to eavesdrop is to tap a wire. Another way is to pick up energy radiated from a wire or equipment (a form of EMI). Years ago, the United States discovered a foreign power doing just that to our embassies. A sensitive antenna in a nearby building was secretly picking up energy radiated from electronic equipment in the embassy. The antenna was receiving EMI much like the radio example discussed earlier. This energy, though, included top-secret and classified data. Businesses also spend millions of dollars each year protecting their secrets, such as encrypting data before it is transmitted.

Fiber optics is a highly secure transmission medium. It does not radiate energy that can be received by a nearby antenna, and it is extremely difficult to tap a fiber. Both government and business consider fiber optics a secure medium.

CONCLUSION

High bandwidth, low loss, and electromagnetic immunity are probably the three most outstanding features of fiber optics. These features dovetail nicely. They allow high-speed data transfer over long distances with little error. A fiber-optic link is capable of transmitting the entire text of a 30-volume encyclopedia over 100 miles in 1 s. The level of error is only one or two incorrect letters during the transmission.

Realize, however, that not all fibers have low loss and high bandwidth. Where loss and high speeds are not critical, as in automobiles, for example, less-expensive fibers work well. In an automobile, the central concern is protecting against noise from sources such as the ignition system. The other advantageous features of fiber optics make it a well-suited transmission medium in many applications.

SUMMARY

- Fiber optics offers many advantages over copper cables.
- Optical fibers offer greater bandwidth. Optical fibers offer lower losses.

- High bandwidth and low losses mean greater repeater spacing in long-distance systems.
- Fibers offer excellent protection against EMI.
- Fibers are a secure transmission medium.
- Fibers are smaller and lighter than comparable copper cables.
- Since fibers do not carry electricity and cannot spark or cause fires, they are safe even in hostile environments.
- Fibers are used for different reasons, such as low loss, high bandwidth, security, and EMI immunity.

 Review Questions

1. List six advantages of fiber optics. Give an example of each.
2. Give three examples of rates and capacities in the North American Digital Telephone Hierarchy.
3. Which of the following is the most important reason for large bandwidth in optical fibers:
   A. High-speed, high-capacity transmission
   B. Secure transmission
   C. Fewer repeaters
   D. Immunity from electromagnetic interference
4. Does loss increase, decrease, or stay the same as the signal frequency increases in a copper cable? In an optical cable?
5. What is the name of the phenomenon that causes ghosts on your television, crackles on your radio, and other malfunctions in electronic equipment? Why does an optical fiber not contribute to this phenomenon?
6. For a 200-kHz signal traveling 50 m through a factory, which of the following factors are of great importance? Of little importance? Explain your reasons.
   A. Bandwidth
   B. Low loss
   C. EMI immunity
   D. Safety
   E. Error-free transmission

7. The original interest in optical fibers as a communication medium arose because
   - **A.** They used less space than a coaxial cable.
   - **B.** They were less expensive than copper.
   - **C.** They could carry the high frequencies of laser light with little loss.
   - **D.** Their electromagnetic immunity, security, and compatibility with digital techniques made them of use to the military.

8. (True/False) Information-carrying capacity depends on the physical size of the medium carrying the capacity.

chapter four

LIGHT

L ight is electromagnetic energy, as are radio waves, radar, television and radio signals, x-rays, and electronic digital pulses. *Electromagnetic energy* is radiant energy that
travels through free space at about 300,000 km/s or 186,000 miles/s. An electromagnetic wave consists of oscillating electric and magnetic fields at right angles to each
other and to the direction of propagation. Thus, an electromagnetic wave is usually
depicted as a sine wave, as shown in Figure 4-1.

The main distinction between different waves lies in their frequency or wavelength.
Frequency, of course, defines the number of sine-wave cycles per second and is expressed
in hertz. *Wavelength* is the distance between the same points on two consecutive waves (or
it is the distance a wave travels in a single cycle). Wavelength and frequency are related.
Wavelength (lambda) equals the velocity of the wave (v) divided by its frequency (f):

$$\lambda = \frac{v}{f}$$

In free space or air, the velocity of an electromagnetic wave is the speed of light.

The equation clearly shows that the higher the frequency, the shorter the wavelength.
For example, the 60-Hz power delivered to your house has a wavelength of 3100 miles. A
55.25-MHz signal, which carries the picture for channel 2 on television, has a wavelength
of 17.8 ft. Deep red light has a frequency of 430 THz (430×10^{12} Hz) and a wavelength
of only 700 nm (nanometers or billionths of a meter).

In electronics, we customarily talk in terms of frequency. In fiber optics, however, light
is described by wavelengths. Remember, though, that frequency and wavelength are
inversely related.

FIGURE 4-1

An electromag-
netic wave

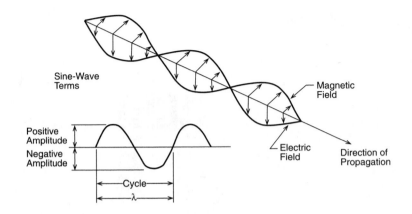

THE ELECTROMAGNETIC SPECTRUM

Electromagnetic energy exists in a continuous spectrum of frequencies from subsonic energy through radio waves, microwaves, γ (gamma) rays, and beyond. (Electromagnetic energy exists in the subsonic range of the spectrum; however, sound itself is not electromagnetic energy but a vibration of air molecules.) Figure 4-2 shows the spectrum. Notice that the radio frequencies most commonly used for communication are well below those of light.

Thus, light is electromagnetic energy with a *higher* frequency and *shorter* wavelength than radio waves. The figure also shows that the visible light we can see is only a small part of the light spectrum. Visible light has wavelengths from 380 nm for deep violet to 750 nm for deep red. Infrared light has longer wavelengths (lower frequencies) than visible light, whereas ultraviolet light has shorter ones. Most fiber-optic systems use infrared light between 800 and 1500 nm because glass fibers carry infrared light more efficiently than visible light.

The high frequencies of light have made it of such interest to communications engineers. As we saw in Chapter 2, a higher-frequency carrier means greater information-carrying capacity. Fiber optics is a method of using this information-carrying potential of light.

WAVES AND PARTICLES

So far, we have described light as an oscillating electromagnetic wave. It is spread out in space, without a definite, discrete location. Physicists once divided all matter into either waves or particles. We usually think of light as a wave and an electron as a particle. Modern physicists, however, have shown that this distinction does not exist. Both light and electrons exhibit wavelike and particle-like characteristics.

FIGURE 4-2
The electromagnetic spectrum (Illustration courtesy of AMP Incorporated)

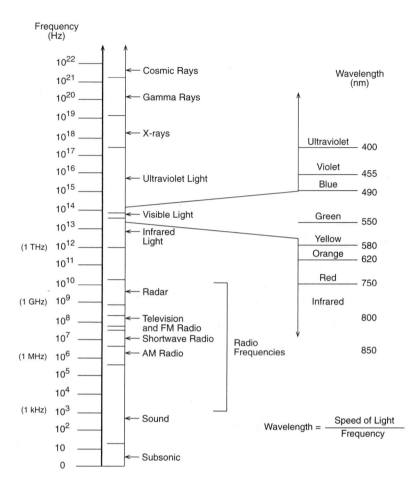

A particle of light is called a *photon,* which is a *quantum* or bundle of energy. A quantum exists in fixed discrete units of energy—you cannot have half a quantum or 5.33 quanta. The amount of energy possessed by a photon depends on its frequency. The amount of energy increases as frequency increases: Higher frequency means more energy. Wavelengths of violet light have more energy than those of red light because they have higher frequencies. The energy *E,* in watts, contained in a photon is

$E = hf$

where *f* is its frequency and *h* is Planck's constant, which is 6.63×10^{-34} J-s (joule-seconds). The equation shows clearly that differences in photon energy are strictly functions of frequency (or wavelength). Photon energy is proportional to frequency. A photon is a quantum of light energy *hf*.

Here are some approximate energy levels for different high-frequency wavelengths. Notice that the higher the frequency, the more energy contained in the quantum.

Infrared light (10^{13} Hz): 6.63×10^{-20} J
Visible light (10^{14} Hz): 6.63×10^{-19} J
Ultraviolet light (10^{15} Hz): 6.63×10^{-18} J
X-rays (10^{18} Hz): 6.63×10^{-15} J

The photon is actually a strange particle, for it has zero rest mass. If it is not in motion, it does not exist! In this sense, it is not a particle as, say, marbles, stones, or ink drops are particles. It is a bundle of energy that acts like a particle.

Treating light as both a wave and a particle aids in investigation of fiber optics. We switch easily between the two descriptions, depending on our needs. For example, many characteristics of optical fibers vary with wavelength, so the wave description is used. On the other hand, the emission of light by a source or its absorption by a detector is best treated by particle theory. The description of a detector speaks of light photons striking the detector and being absorbed. This absorption provides the energy required to set electrons flowing as current. A light-emitting diode (LED) operates because its electrons give up energy as photons. The exact energy of the photon determines the wavelength of the emitted light.

LIGHT RAYS AND GEOMETRIC OPTICS

The simplest way to view light in fiber optics is by ray theory. The light is treated as a simple ray, shown by a line. An arrow on the line shows the direction of propagation. The movement of light through the fiber-optic system can be analyzed with simple geometry. This approach not only simplifies analysis, but it also makes the operation of an optical fiber simple to understand.

REFLECTION AND REFRACTION

What is commonly called the *speed of light* is actually the velocity of electromagnetic energy in a vacuum such as space. Light travels at slower velocities in other materials such as glass. Light traveling from one material to another changes speed, which, because of wave motion, results in light changing its direction of travel. This deflection of light is called *refraction*. In addition, different wavelengths of light travel at different speeds in the *same* material. The variation of velocity with wavelength plays an important role in fiber optics.

FIGURE 4-3
Refraction

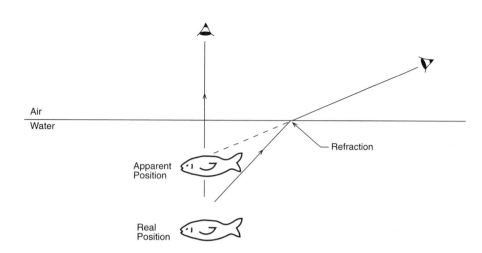

People who fish are quite familiar with refraction, since it distorts the apparent position of a fish lying stationary underwater (Figure 4-3). If one stands on a pier and looks directly down on the fish, the light is not refracted so that the fish is where it appears to be. If the fish is viewed by looking into the water at an angle, refraction of light occurs. What appears to be straight line from the fish to the eye is actually a line with a bend where the light passes from water into air and is refracted. As a result, the fish is actually deeper in the water than it appears to be.

The prism in Figure 4-4 also demonstrates refraction. White light entering the prism contains all colors. The prism refracts the light, and it changes speed as it enters the prism. Because each color or frequency changes speed differently, each is refracted differently. Red light deviates the least and travels the fastest. Violet light deviates the most and travels the slowest. The light emerges from the prism divided into the colors of the rainbow. Notice that refraction occurs at the entrance and at the exit of the prism.

FIGURE 4-4
Refraction and a prism (Illustration courtesy of AMP Incorporated)

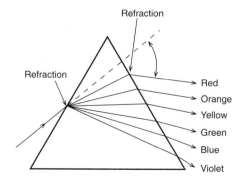

FIGURE 4-5
Indices of refraction for various materials

MATERIAL	INDEX (n)	LIGHT VELOCITY (km/s)
Vacuum	1.0	300,000
Air	1.0003 (1)	300,000
Water	1.33	225,000
Fused quartz	1.46	205,000
Glass	1.5	200,000
Diamond	2.5	120,000

The index of refraction, symbolized by n, is a dimensionless number expressing the ratio of the velocity of light c in free space to its velocity v in a specific material:

$$n = \frac{c}{v}$$

Figure 4-5 lists some representative indices of refractions for selected materials as well as the approximate speed of light through the materials.

Of particular importance to fiber optics is that the glass's index of refraction can be changed by controlling its composition.

The amount that a ray of light is refracted depends on the refractive indices of the two materials. But before looking at the mechanics of refraction, we must first define some terms vital to our discussion. Figure 4-6 illustrates several important terms to understand light and its refraction.

FIGURE 4-6
Angles of incidence and refraction (Illustration courtesy of AMP Incorporated)

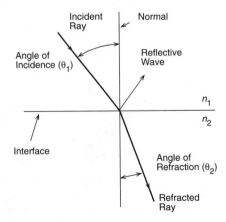

FIGURE 4-7
Reflection
(Illustration
courtesy of AMP
Incorporated)

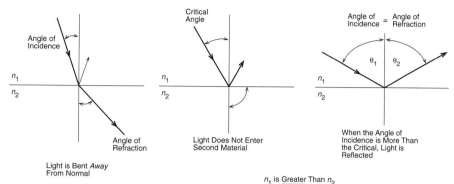

The *normal* is an imaginary line perpendicular to the interface of the two materials.
- The *angle of incidence* is the angle between the incident ray and the normal.
- The *angle of refraction* is the angle between the refracted ray and the normal.

Light passing from a lower refractive index to a higher one is bent *toward* the normal. But light going from a higher index to a lower one refracts *away* from the normal, as shown in Figure 4-7. As the angle of incidence increases, the angle of refraction approaches 90° to the normal. The angle of incidence that yields an angle of refraction of 90° is the *critical* angle. If the angle of incidence increases past the critical angle, the light is totally reflected back into the first material so that it does not enter the second material. The angles of incidence and reflection are equal.

FRESNEL REFLECTIONS

Even when light passes from one index to another, a small portion is always reflected back into the first material. These reflections are *Fresnel reflections*. A greater difference in the indices of the materials results in a greater portion of the light reflecting. Fresnel reflection (ρ) at the boundary between air and another equals

$$\rho = \left(\frac{n-1}{n+1} \right)^2$$

In decibels, this loss of transmitted light is

$$dB = 10 \log_{10} (1 - \rho)$$

For light passing from air to glass (with $n = 1.5$ for glass), Fresnel reflection is about 0.17 dB. This figure will vary somewhat as the composition of the glass varies. Since such losses occur when light enters or exits an optical fiber, the loss in joining one fiber to another is 0.34 dB. A Fresnel reflection occurs when the light passes from the first fiber into the air gap separating the two fibers. A second Fresnel reflection occurs when the light passes from the air into the second fiber. Fresnel reflection is the same regardless of the order of materials through which the light passes; in other words, Fresnel reflection is the same whether light passes from glass to air or from air to glass.

SNELL'S LAW

Snell's law states the relationship between the incident and refracted rays:

$$n_1 \sin \theta_1 = n_2 \sin \theta_2$$

where θ_1 and θ_2 are defined in Figures 4-6 and 4-7.

The law shows that the angles depend on the refracted indices of the two materials. Knowing any three of the values, of course, allows us to calculate the fourth through simple rearrangement of the equation.

The critical angle of incidence θ_c, where $\theta_2 = 90°$, is

$$\theta_c = \arcsin \left(\frac{n_2}{n_1} \right)$$

At angles greater than θ_c, the light is reflected. Because reflected light means that n_1 and n_2 are equal (since they are in the same material), θ_1 and θ_2 are also equal. The angles of incidence and reflection are equal. These simple principles of refraction and reflection form the basis of light propagation through an optical fiber.

A PRACTICAL EXAMPLE

Figure 4-8 shows an example of reflection that has practical application in fiber optics. Assume we have two layers of glass, as shown in Figure 4-8A. The first layer, n_1, has a refractive index of 1.48; the second, 1.46. These values are typical for optical fibers. Using Snell's law, we can calculate the critical angle:

$$\theta_c = \arcsin \left[\frac{1.46}{1.48} \right]$$
$$= \arcsin (0.9864)$$
$$= 80.6°$$

FIGURE 4-8

A practical example of reflection

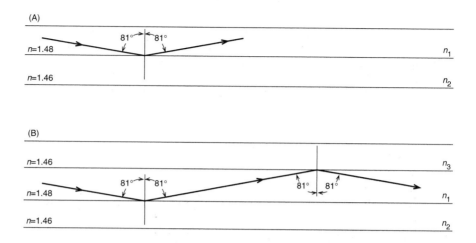

Light striking the boundary between n_1 and n_2 at an angle greater than 80.6° from the normal reflects back into the first material. Again, the angle of incidence equals the angle of reflection.

Figure 4-8B carries the example a step further. Assume that a third layer of glass—labeled n_3—with the same refractive index as n_2 is placed on top of material n_1. Material n_1 is sandwiched between the n_2 and n_3 materials. We again have the same boundary condition as before. The reflected ray, however, now becomes the incident ray at the new boundary. The critical angle remains 80.6°. Thus all conditions are the same as in the first reflection. As a result, the light is again reflected back into the first material. The light reflected from n_3 becomes the incident ray again for n_2. The situation is the same. We have trapped the light between the two layers n_2 and n_3. As long as the angle of incidence is greater than 80.6°, the light is reflected back into the first material. Snell's law shows that, for this ideal example, this will always hold true. The light will continue traveling through the first material by total internal reflection.

The same principle explains the operation of an optical fiber. The main difference is that the fiber uses a circular configuration, such that n_2 surrounds n_1. The next chapter looks at light propagation in a fiber in detail.

SUMMARY

- Light is electromagnetic energy with a higher frequency and shorter wavelength than radio waves.
- Light has both wavelike and particlelike characteristics.
- When light meets a boundary separating materials of different refractive indices, it is either refracted or reflected.

- Fresnel reflections occur regardless of the angle of incidence.
- Snell's law describes the relationship between incident and reflected light.

Review Questions

1. Describe the frequency and wavelength of light in relation to radio frequencies.
2. What is the name given to a particle of light?
3. Sketch a light ray being refracted. Label the incident ray, the refracted ray, and the normal.
4. Sketch a light ray being reflected. Label the incident ray, the reflected ray, and the normal.
5. If an incident ray passing from air into water has a 75° angle with respect to the normal, what is the angle of refracted ray?
6. What is the wavelength of a 300-MHz electromagnetic wave in free space? A 250-kHz wave? A 2-GHz wave?
7. Define Fresnel reflection.
8. Define critical angle.
9. Does light travel faster in air or in glass?

PART TWO

FIBER-OPTIC COMPONENTS

chapter five

THE OPTICAL FIBER

The last chapter showed the characteristics of light propagation most important to your understanding of fiber optics. We saw that the reflection or refraction of light depends on the indices of refraction of the two media and on the angle at which light strikes the interface. The optical fiber works on these principles. Once light begins to reflect down the fiber, it will continue to do so under normal circumstances. The purpose of this chapter is to describe the propagation of light through the various types of optical fibers. The next chapter further examines the properties of fibers.

Keep in mind the distinction between the optical fiber and the fiber-optic cable. The *optical fiber* is the signal-carrying member, similar in function to the metallic conductor in a wire. But the fiber is usually cabled—that is, placed in a protective covering that keeps the fiber safe from environmental and mechanical damage. This chapter deals specifically with the optical fiber itself.

BASIC FIBER CONSTRUCTION

The optical fiber has two concentric layers called the *core* and the *cladding*. The inner core is the light-carrying part. The surrounding cladding provides the difference in refractive index that allows total internal reflection of light through the core. The index of the cladding is less than 1% lower than that of the core. Typical values, for example, are a core index of 1.47 and a cladding index of 1.46. Fiber manufacturers must carefully control this difference to obtain desired fiber characteristics.

Fibers have an additional coating around the cladding. The coating, which is usually one or more layers of polymer, protects the core and cladding from shocks that might affect their optical or physical properties. The coating has no optical properties affecting the propagation of light within the fiber. This coating, then, is a shock absorber.

Figure 5-1 shows the idea of light traveling through a fiber. Light injected into the fiber and striking the core-to-cladding interface at greater than the critical angle reflects back into the core. Since the angles of incidence and reflection are equal, the reflected light will again be reflected. The light will continue zigzagging down the length of the fiber.

Light, however, striking the interface at less than the critical angle passes into the cladding, where it is lost over distance. The cladding is usually inefficient as a light carrier, and light in the cladding becomes attenuated fairly rapidly.

Notice also, in Figure 5-1, that the light is also refracted as it passes from air into the fiber. Thereafter, its propagation is governed by the indices of the core and cladding and by Snell's law.

Such total internal reflection forms the basis of light propagation through a simple optical fiber. This analysis, however, considers only *meridional rays*—those that pass through the fiber axis each time they are reflected. Other rays, called *skew rays,* travel down the fiber without passing through the axis. The path of a skew ray is typically helical, wrapping around and around the central axis. Fortunately, skew rays are ignored in most fiber-optic analyses.

The specific characteristics of light propagation through a fiber depend on many factors, including

- The size of the fiber
- The composition of the fiber
- The light injected into the fiber

An understanding of the interplay between these properties will clarify many aspects of fiber optics.

Fibers themselves have exceedingly small diameters. Figure 5-2 shows cross sections of the core and cladding diameters of four commonly used fibers. The diameters of the core and cladding are provided on the following page.

FIGURE 5-1
Total internal reflection in an optical fiber (Illustration courtesy of AMP Incorporated)

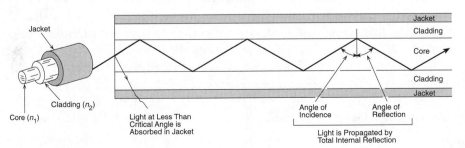

FIGURE 5-2
Typical core
and cladding
diameters

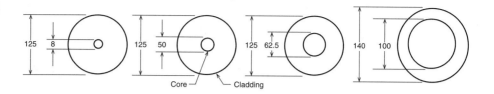

Core (μm)	Cladding (μm)
8	125
50	125
62.5	125
100	140

To realize how small these sizes are, note that human hair has a diameter of about 100 μm. Fiber sizes are usually expressed by first giving the core size, followed by the cladding size: thus, 50/125 means a core diameter of 50 μm and a cladding diameter of 125 μm; 100/140 means a 100-μm core and a 140-μm cladding. Through these small sizes are sent thousands of telephone conversations.

FIBER CLASSIFICATION

Optical fibers are classified in two ways. One way is by their material makeup:

- Glass fibers have a glass core and glass cladding. Since they are by far the most widely used, most discussion in this book centers on glass fibers. The glass used in fibers is ultrapure, ultratransparent silicon dioxide or fused quartz. If seawater were as clear as a fiber, you could see to the bottom of the deepest ocean trench, the 35,839-foot-deep Marianas Trench in the Pacific. Impurities are purposely added to the pure glass to achieve the desired index of refraction. Germanium or phosphorus, for example, increases the index. Boron or fluorine decreases the index. Other impurities not removed when the glass is purified also remain. These, too, affect fiber properties by increasing attenuation by scattering or absorbing light.
- Plastic-clad silica (PCS) fibers have a glass core and plastic cladding. Their performance is not as good as all-glass fibers.
- Plastic fibers have a plastic core and plastic cladding. Compared with other fibers, plastic fibers are limited in loss and bandwidth. Their very low cost and easy use, however, make them attractive in applications where high bandwidth or low loss is not a concern. Their electromagnetic immunity and security

allow plastic fibers to be beneficially used. Plastic and PCS fibers do not have the buffer coating surrounding the cladding.

The second way to classify fibers is by the refractive index of the core and the modes that the fiber propagates. Figure 5-3, which depicts the differences in fibers classified this way, shows three important ideas about fibers.

First, it shows the difference between the input pulse injected into a fiber and the output pulse emerging from the fiber. The decrease in height of the pulse shows loss of signal power. The broadening in width limits the fiber's bandwidth or signal-carrying capacity. Second, it shows the path followed by light rays as they travel down the fiber. Third, it shows the relative index of refraction of the core and cladding for each type of fiber. The significance of these ideas will become apparent as we examine each type of fiber.

MODES

Mode is a mathematical and physical concept describing the propagation of electromagnetic waves through media. In its mathematical form, mode theory derives from Maxwell's equations. James Clerk Maxwell, a Scottish physicist in the last century, first gave mathematical expression to the relationship between electric and magnetic energy. He showed that they were both a single form of electromagnetic energy, not two different

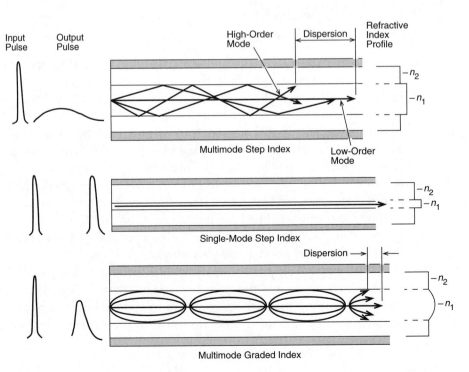

FIGURE 5-3
Types of fiber propagation (Illustration courtesy of AMP Incorporated)

forms as was then believed. His equations also showed that the propagation of this energy followed strict rules. Maxwell's equations form the basis of electromagnetic theory.

A mode is an allowed solution to Maxwell's equations. For purposes of this book, however, a mode is simply a path that a light ray can follow in traveling down a fiber. The number of modes supported by a fiber ranges from 1 to over 100,000. Thus, a fiber provides a path of travel for one or thousands of light rays, depending on its size and properties.

REFRACTIVE INDEX PROFILE

The *refractive index profile* describes the relation between the indices of the core and cladding. Two main relationships exist: step index and graded index. The step-index fiber has a core with a uniform index throughout. The profile shows a sharp step at the junction of the core and cladding. In contrast, the graded index has a nonuniform core. The index is highest at the center and gradually decreases until it matches that of the cladding. There is no sharp break between the core and the cladding.

By this classification, there are three types of fibers (whose names are often shortened by a prevalent characteristic):

1. Multimode step-index fiber (commonly called step-index fiber)
2. Multimode graded-index fiber (graded-index fiber)
3. Single-mode step-index fiber (single-mode fiber)

The characteristics of each type have important bearing on its suitability for particular applications. As we progress through this chapter and the next, the importance of each type will become apparent.

STEP-INDEX FIBER

The multimode step-index fiber is the simplest type. It has a core diameter from 100 to 970 μm, and it includes glass, PCS, and plastic constructions. As such, the step-index fiber is the most wide ranging, although not the most efficient in having high bandwidth and low losses.

Since light reflects at different angles for different paths (or modes), the path lengths of different modes are different. Thus, different rays take a shorter or longer time to travel the length of the fiber. The ray that goes straight down the center of the core without reflecting arrives at the other end first. Other rays arrive later. Thus, light entering the fiber at the same time exits the other end at different times. The light has spread out in time.

This spreading of an optical pulse is called *modal dispersion*. A pulse of light that began as a tightly and precisely defined shape has dispersed—spread over time. Dispersion describes the spreading of light by various mechanisms. Modal dispersion is that type of dispersion that results from the varying path lengths of different modes in a fiber.

You can image three race cars all traveling the same speed. The first race car follows a straight path—equal to the lowest order mode that does not reflect as it travels. The second car follows the longest path—equal to the highest order mode. While its speed does not change from that of the first car, it must travel continuously back and forth through many curves. The third car follows an intermediate path. If all three cars begin at the same time and travel to a finish line one mile away, they obviously will arrive there at different times. The same holds true for a pulse of light injected into a fiber. Different rays will follow different paths and so arrive at different times.

Typical modal dispersion figures for step-index fibers are 15 to 30 ns/km. This means that when rays of light enter a fiber at the same time, the ray following the longest path will arrive at the other end of a 1-km-long fiber 15 to 30 ns after the ray following the shortest path.

Fifteen to 30 billionths of a second may not seem like much, but dispersion is the main limiting factor on a fiber's bandwidth. Pulse spreading results in a pulse overlapping adjacent pulses, as shown in Figure 5-4. Eventually, the pulses will merge so that one pulse cannot be distinguished from another. The information contained in the pulse is lost. Reducing dispersion increases fiber bandwidth.

GRADED-INDEX FIBER

One way to reduce modal dispersion is to use graded-index fibers. Here the core has numerous concentric layers of glass, somewhat like the annular rings of a tree. Each successive layer outward from the central axis of the core has a lower index of refraction. Figure 5-5 shows the core's structure.

Light, remember, travels faster in a lower index of refraction. So the further the light is from the center axis, the greater its speed. Each layer of the core refracts the light. Instead of being sharply reflected as it is in a step-index fiber, the light is now bent or continually refracted in an almost sinusoidal pattern. Those rays that follow the longest path by traveling near the outside of the core have a faster average velocity. The light

FIGURE 5-4
Pulse spreading

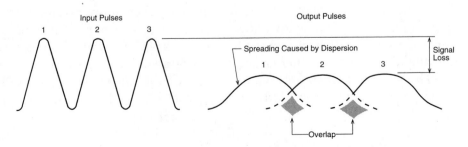

FIGURE 5-5
Concentric lay-
ers of lower
refractive index
in a graded-
index core
(Courtesy of
AT&T/Lucent
Technologies
Bell Labs)

traveling near the center of the core has the slowest average velocity. As a result, all rays tend to reach the end of the fiber at the same time. The graded index reduces modal dispersion to 1 ns/km or less.

Popular graded-index fibers have core diameters of 50 or 62.5 μm and a cladding diameter of 125 μm. The fiber is popular in applications requiring a wide bandwidth, especially telecommunications, local area networks, computers, and similar uses.

SINGLE-MODE FIBER

Another way to reduce modal dispersion is to reduce the core's diameter until the fiber propagates only one mode efficiently. The single-mode fiber has an exceedingly small core diameter of only 5 to 10 μm. Standard cladding diameter is 125 μm. This cladding diameter was chosen for three reasons:

1. The cladding must be about 10 times thicker than the core in a single-mode fiber. For a fiber with an 8- or 9-μm core, the cladding should be at least 80 μm.
2. It is the same size as graded-index fibers, which promotes size standardization.
3. It promotes easy handling because it makes the fiber less fragile and because the diameter is reasonably large so that it can be handled by technicians.

Since this fiber carries only one mode, modal dispersion does not exist.

Single-mode fibers easily have a potential bandwidth of 50 to 100 GHz-km. Present fibers have a bandwidth of several gigahertz and allow transmission of tens of kilometers. As of early 1985, the largest commercially available fiber-optic systems were digital telephone transmission systems operating around 417 Mbps. These systems carried 6048 simultaneous telephone calls over a single-mode fiber a distance of 35 km without a repeater. By the end of 1992, capacities had grown fourfold to 10 Gbps and 130,000 voice channels. More importantly, fibers now carry multiple wavelengths to increase capacities to hundreds of gigabits per second.

An important aspect of this growth is that the increase results from changing the electronics, not the single-mode fibers, at either end of the system. The capacity of a single-mode system is limited by the capabilities of the electronics, not of the fiber. One advantage of single-mode fibers is that once they are installed, the system's capacity can be increased as newer, higher-capacity transmission electronics becomes available. This capability saves the high cost of installing a new transmission medium to obtain increased performance and allows cost-effective increases in data rates.

The point at which a single-mode fiber propagates only one mode depends on the wavelength of light carried. A wavelength of 850 nm results in multimode operation. As the wavelength is increased, the fiber carries fewer and fewer modes until only one remains. Single-mode operation begins when the wavelength approaches the core diameter. At 1300 nm, for example, the fiber permits only one mode. It becomes a single-mode fiber.

Different fiber designs have a specific wavelength, called the *cutoff* wavelength, above which it carries only one mode. A fiber designed for single-mode operation at 1300 nm has a cutoff wavelength of around 1200 nm.

The operation of a single-mode fiber is slightly more complex than simply a ray traveling down the core. Geometric optics using light rays is not as appropriate for these fibers because it obscures how optical energy is distributed within the fiber. Some of the optical energy of the mode travels in the cladding, as shown in Figure 5-6. Therefore, the diameter of the light appearing at the end of the fiber is larger than the core diameter. *Mode field diameter* is the term used to define this diameter of optical energy. Although optical energy is confined to the core in a multimode fiber, it is not so confined in a single-mode fiber. It is usually more important to know the mode field diameter than the core diameter.

The difference in propagation of light in a single-mode fiber points to another difference between single-mode and multimode fibers. Because optical energy in a single-mode fiber travels in the cladding as well as in the core, the cladding must be a more efficient carrier of energy. In a multimode fiber, the light transmission characteristics of the cladding are basically unimportant. Indeed, because cladding modes are not desirable, a cladding with inefficient transmission characteristics can be tolerated. This situation does not hold for a single-mode fiber.

FIGURE 5-6
Optical power in
multimode and
single-mode
fibers (Courtesy
of Corning Glass
Works)

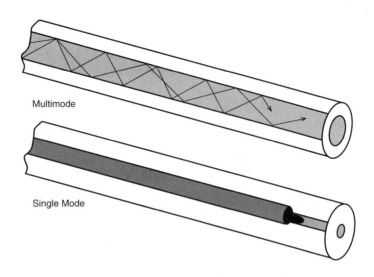

Multimode

Single Mode

DISPERSION-SHIFTED
SINGLE-MODE FIBERS

There are three types of single-mode optical fibers usually found in typical applications for telecommunications and data networking. Beyond standard single-mode fibers, there are also dispersion-shifted (DS) fibers and nonzero-dispersion-shifted (NZ-DS) fibers. The purpose of these fibers is to reduce dispersion in the transmission window having the lowest attenuation. Normally, attenuation is lowest in the 1550-nm windows and dispersion is lowest at 1300 nm. Dispersion shifting creates a fiber that shifts the low-dispersion region to the 1550-nm region. This shifting of dispersion results in a fiber suited to the highest data rates and longest transmission distances. In a standard single-mode fiber, the points of lowest loss and highest bandwidth do not coincide. Dispersion shifting bring them closer together. The next chapter discusses dispersion in detail and will describe DS and NZ-DS fibers further.

SHORT-WAVELENGTH
SINGLE-MODE FIBERS

A single-mode fiber can be constructed with a shorter cutoff wavelength. Some fibers have been designed with a cutoff wavelength of 570 nm for operation at 633 nm (which is visible red light). The core is quite small, less than 4 μm. Another fiber has a cutoff wavelength of 1000 nm, a recommended operating frequency of 1060 nm, and a core diameter of under 6 μm. These fibers are intended for specialized telecommunications, computer, and sensor applications. They are not a replacement for standard long-distance telecommunication single-mode fibers operating at the longer wavelengths of 1300 and 1550 nm. For one thing, their attenuation levels are several times higher—up to 10 dB/km for the 633-nm fiber—making them unsuited for longer distances.

PLASTIC OPTICAL FIBERS

Although most of the discussion in this book involves glass fibers, plastic optical fibers should not be overlooked. The main attraction of a plastic fiber is price—components are less expensive. This includes not only the optical-fiber cable, but sources, detectors, and connectors as well. The drawback is performance—plastic optical fibers do not have the high bandwidth or low attenuation of glass fibers.

Traditionally, plastic fibers have been step-index components. A high-end application is 50 Mbps over 100 meters. The fibers have a large core and very thin cladding. Typical sizes are 480/500, 735/750, and 980/1000 μm, although tolerances in plastic fibers are much looser than with glass fibers.

More recently, graded-index plastic fibers were introduced. These offer data rates of up to 3 Gbps over 100 meters. As such, they are being promoted as a low-cost alternative to both glass fibers and copper cables in premises cabling, local area networks, home networks, and similar applications where distances are limited. The high bandwidth allows them to meet the performance requirements of typical networks. For example, most network applications limit distances between devices to 100 meters.

Plastic fibers are typically made of polymethylmethacrylate (PMMA), which has a low-attenuation band at 650 nm in the visible spectrum. Attenuation is about 150 dB/km or 15 dB over 100 meters. Standard PMMA fibers are not suited for operation at the common 850, 1300, and 1550 nm regions used in telecommunications and data networking. A fluorinated plastic fiber can achieve acceptably low losses at 850 and 1300 nm.

Plastic fibers have several distinct features that make them attractive in cost-sensitive applications. Since components cost less, installations are less expensive.

Plastic fibers use red light in the 650-nm range. Visible light aids diagnostics and troubleshooting since the presence of light in the system is easy to see. In addition, the optical power and output patterns do not pose the safety concerns associated with infrared laser light.

Plastic fibers are quite rugged, with a tight bend radius and the ability to withstand abuse. Their EMI immunity makes them attractive in noisy environments.

Plastic fibers are easy for a technician to work with. Connectors are quickly and easily applied, typically in a minute or less.

Because of the low cost, good performance, and ruggedness of optical fibers, they find use in such applications as automobiles, industrial automation, music systems, and other consumer electronics. The Japanese have created standards for using plastic fibers in home electronics systems to interconnect such equipment as digital audio tape players, MIDI equipment, and compact disk players. Plastic fibers have been recommended for home wiring and computer networks.

HOW MANY MODES ARE THERE?

We have seen that the number of modes supported by a fiber in part determines its information-carrying capacity. Modal dispersion, which causes pulse spreading and overlapping, limits the data rate that a fiber can support. We have also seen that dispersion depends on wavelength and core diameter.

The V number, or normalized frequency, is a fiber parameter that takes into account the core diameter, wavelength propagated, and fiber NA (a fiber property described in the next chapter):

$$V = \frac{2\pi d}{\lambda} \text{(NA)}$$

From the V number, the number of modes in a fiber can be calculated.

For a simple step-index fiber, the number of modes can be approximated by

$$N = \frac{V^2}{2}$$

For a graded-index fiber, the number of modes can be approximated by

$$N = \frac{V^2}{4}$$

The equations demonstrate that the number of modes is determined by core diameter, fiber NA, and the wavelength propagated. The number of modes in a graded-index fiber is about half that of a step-index fiber having the same diameter and NA. A fiber with a 50-μm core supports over 1000 modes.

When the V number of a step-index fiber becomes 2.405, the fiber supports a single mode. The V number can be decreased by decreasing the core diameter, by increasing the operating wavelength, or by decreasing the NA. Thus, single-mode operation in a fiber can be obtained by suitably adjusting these characteristics.

Figure 5-7 shows the number of modes contained in three different common fiber sizes operating at two different wavelengths. In the same fiber, the longer wavelength of 1300 nm travels in half as many modes as the shorter 850-nm wavelength. Similarly, decreasing the core diameter also significantly reduces the number of modes.

FIGURE 5-7
Number of
modes in various
fiber types at two
different wave-
lengths

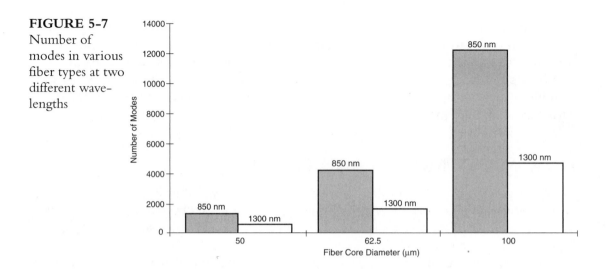

FIBER COMPARISONS

Figure 5-8 shows typical characteristics of various fibers. The meaning of terms such as NA are explained in the next chapter. For now, though, you can see that the performance and physical properties cover a broad range. We use the term "performance" broadly: Better performance means higher bandwidth, higher information-carrying capacity, and lower losses. Other measures of performance, such as safety or low cost, would favor other types of fibers. The table also suggests some generalizations about loss and bandwidth:

- Fiber performance from lowest to highest is as follows:
  –Plastic
  –PCS
  –Step-index glass
  –Graded index
  –Single mode
- A smaller core usually means better performance.
- Glass fibers perform better than plastic ones.

Remember, though, that such generalizations suffer the fault of all generalizations: They do not tell the whole story. A fiber-optic cable must be matched to the requirements of the application. A system needing to transmit only a few thousand bits a second over a couple of meters can use plastic fibers to best advantage. Not only does plastic fiber cost less, but compatible components such as sources, detectors, and connectors also cost less. Using a single-mode fiber in such applications would be like using a Ferrari to go to the corner store. It is common to stress the low losses and high bandwidths of fibers, but not all applications need such performance. All the fiber types have their uses.

FIGURE 5-8
Typical fiber characteristics

FIBER TYPE	CORE DIAMETER (μm)1	CLADDING DIAMETER (μm)	NA	ATTENUATION (db/km) (MAX)					BANDWIDTH (MHz/km) (MAX)
				650	790	850	1300	1550	
Single Mode	3.7	80 or 125		10					
	5.0	85 or 125				2.3			5000 @ 850 nm
	9.3	125	0.13				0.4	0.3	6 ps/km^2
	8.1	125	0.17				0.5	0.25	
Graded Index	50	125	0.20			2.4	0.6	0.5	600 @ 850 nm
									1500 @ 1300 nm
	62.5	125	0.275			3.0	0.7	0.3	200 @ 850 nm
									1000 @ 1300 nm
	100	140	0.29			3.5	1.5	0.9	300 @ 850 nm
									500 @ 1300 nm
Step Index	200	380	0.27			6.0			6 @ 850 nm
	300	440	0.27			6.0			6 @ 850 nm
PCS	200	350	0.30		10				20 @ 790 nm
Plastic	485	500	0.5	240					5 @ 680 nm^3
	735	750	0.5	230					
	980	1000	0.5	220					

1. Mode field diameter given for single-mode fiber; actual core diameter is less.
2. Dispersion per nanometer of source width.
3. Plastic fibers typically are used at distances under 100 m, with data rates up to 50 Mbit/s.

MANUFACTURING OPTICAL FIBER

Optical fibers are made to exacting dimension and compositional requirements. There are three basic steps in making a fiber: the laydown, the consolidation, and the drawing (Figure 5-9).

Fiber manufacturing starts with a preform, which is a large cylinder of glass with the same composition as the final fiber, but at a much larger diameter. A typical preform can be drawn into 10 km of fiber. The first step of making a preform is the laydown. Here, ultrapure vapors containing silica and germania travel through a traversing burner. They react to form soot particles that combine to form the soot preform.

Different methods are used to make the preform. The most common method is the *outside vapor deposition* (OVD) techinique, which deposits the soot on a rotating target rod. The core is deposited first and then the cladding. The vapors used to form the soot are changed as the composition of the preform is changed. For example, the core and cladding require different vapor mixes. After the vapors are deposited, the rod is removed and the preform is placed in a consolidation oven. The oven removes water vapor and sinters the

FIGURE 5-9
Fiber manufac-
turing: preform
laydown and
fiber drawing
(Coutesy of
Corning Glass
Works)

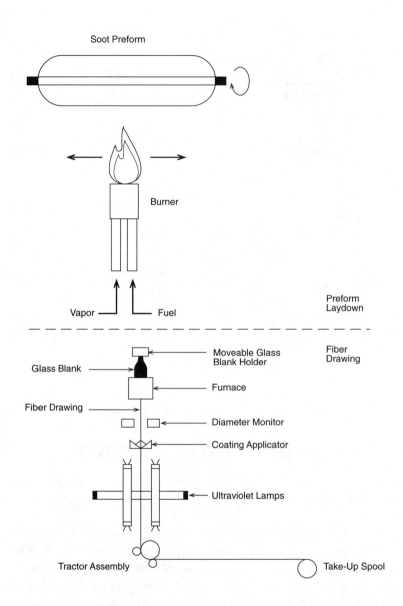

preform in a solid, transparent cylinder. The hole from the rod disappears. Other methods
include:

- *Inside vapor deposition* (IVD), where the soot is deposited inside a hollow silica
  tube. This was the method used by Corning to create the first optical fibers in
  the 1970s. The tube is about 1 meter long and 2.5 cm in diameter (3 feet by 1
  inch).

- *Vapor axial deposition* (VAD), where soot is deposited from the end.
- *Modified chemical vapor deposition* (MCVD), which is a variation of IVD.

Each method has advantages and disadvantages, depending on the type of fiber being made. For example, the OVD method is very good for step-index single-mode fiber. IVD and MCVD are used for high-preformance single-mode fiber.

After the preform is finished, it is drawn into a fiber. The preform is placed in a drawing tower. A furnace, operating at 2500°C, heats the preform until a gob begins to fall. As the gob falls, it pulls a thin fiber behind it. The fiber, continuously drawn through the furnace, has the same relative geometry and composition as the preform, but is in the small diameter of the finished fiber. A tractor assembly at the bottom of the tower ensures that the fiber is drawn at a consistent speed so that the dimensional accuracy is maintained.

As the fiber is drawn, it first passes through a diameter monitor to ensure the cladding diameter remains within specifications. Next the buffer costing is applied and then cured as it passes through ultraviolet lights. Finally, the new fiber is wound onto a spool.

Finished fibers are tested. They are then ready for cabling. Cabling is often done by a company other than the one who made the fiber.

SUMMARY

- The three types of fibers are step index, graded index, and single mode.
- Dispersion is one factor limiting fiber performance. One motive of fiber design is to reduce dispersion by grading the index or by using a single-mode design.
- Core diameter gives a rough estimate of fiber performance: the smaller the core, the higher the bandwidth and the lower the loss.
- Fibers come in a variety of performances to suit different application needs.
- Mode field diameter is the diameter of optical energy carried in a single-mode fiber.
- Single-mode fibers use both step-index and more complex profiles.

 Review Questions

1. What are the two main parts of an optical fiber?
2. What are the three types of fibers according to material composition?

3. Light travels down a step-index fiber by what principle?
4. What usually happens to the performance of a fiber if the core diameter is reduced? If it is enlarged?
5. What is the path called that is followed by optical energy down a fiber?
6. What is the name of a fiber that allows only one path for the light down the fiber?
7. What is the name of a fiber with a core whose refractive index varies?
8. Is the refractive index of the core higher or lower than that of the cladding?
9. (True/False) Modal dispersion is constant at all wavelengths.
10. (True/False) One advantage of the graded-index fiber is that the wavelengths of low loss and minimum dispersion always coincide.

chapter six

FIBER CHARACTERISTICS

This chapter examines the characteristics of optical fibers most important to users and designers. The chapter expands on the characteristics discussed in previous chapters and introduces new ones.

DISPERSION

As discussed in the last chapter, *dispersion* is the spreading of a light pulse as it travels down the length of an optical fiber. Dispersion limits the bandwidth or information-carrying capacity of a fiber. The bit rate must be low enough to ensure that pulses do not overlap. A lower bit rate means that the pulses are farther apart and, therefore, that greater dispersion can be tolerated. There are four main types of dispersion:

1. Modal dispersion
2. Material dispersion
3. Waveguide dispersion
4. Polarization mode dispersion

MODAL DISPERSION

As shown in the last chapter, modal dispersion occurs only in multimode fibers. It arises because rays follow different paths through the fiber and consequently arrive at the other end of the fiber at different times. Modal dispersion can be reduced in three ways:

1. Use a smaller core diameter, which allows fewer modes. A 100-μm-diameter core allows fewer paths than a 200-μm-diameter core.
2. Use a graded-index fiber so that the light rays that follow longer paths also travel at a faster average velocity and thereby arrive at the other end of the fiber at nearly the same time as rays that follow shorter paths.
3. Use a single-mode fiber, which permits no modal dispersion.

MATERIAL DISPERSION

Different wavelengths (colors) also travel at different velocities through a fiber, even in the same mode. Earlier, we saw that the index of refraction n is equal to

$$n = \frac{c}{v}$$

where c is the speed of light in a vacuum and v is the speed of the same wavelength in a material.

Each wavelength, though, travels at different speeds through a material, so the value of v in the equation changes for each wavelength. Thus, the index of refraction changes according to the wavelength. Dispersion from this phenomenon is called either *material dispersion* (since it arises from the material properties of the fiber) or *chromatic dispersion* (since it arises from the different wavelengths traveling through the fiber). The amount of dispersion depends on two factors.

1. The range of wavelengths injected into the fiber. A source does not emit a single wavelength—it emits several. The range of wavelengths, expressed in nanometers, is the spectral width of the source. An LED can have a spectral width in the range of 35 nm to well over 100 nm. A laser diode has a much narrower spectrum, from 0.1 nm to 3 nm.
2. The nominal operating wavelength of the source. Around 850 nm, longer "reddish" wavelengths travel faster than shorter ("bluish") wavelengths. An 860-nm ray travels through glass faster than an 840-nm ray. At 1550 nm, the situation is reversed. The shorter wavelength travels faster than longer ones: a 1560-nm ray travels slower than a 1540-nm wave. At some point, a crossover must occur where the bluish and reddish wavelengths travel at the same speed. In a standard single-mode fiber, this point—called the zero dispersion point—is around 1300 nm. Figure 6-1 shows this idea. The length of the arrows suggests the speed of the wavelength; hence, a longer arrow means that the wave is traveling faster and arriving earlier.

Material dispersion is of greater concern in single-mode systems. In a multimode system, modal dispersion is usually significant enough that material dispersion is not a factor.

FIGURE 6-1

Material dispersion and wavelength

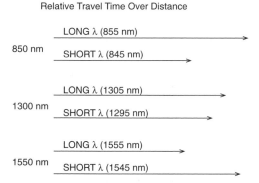

Relative Travel Time Over Distance

In many cases, designers are not even concerned with material dispersion because speeds are too low or distances are too short.

A standard single-mode fiber has the lowest material dispersion at 1300 nm and the lowest loss at 1550 nm. In other words, it has the highest information-carrying capacity at 1300 nm and the longer transmission distance at 1550 nm. This obviously presents a trade-off, leaving a designer to choose one wavelength for higher data rates and another for longer transmission distances. Dispersion is about five times higher at 1550 nm than at 1300 nm, while attenuation is 0.2 dB lower. In a standard single-mode fiber, material dispersion at 1550 nm is about 16 ps per nanometer of source width per kilometer.

A *dispersion-shifted fiber* attempts to give the designer the best of both worlds. Low loss and high bandwidth at the same optical wavelength. The zero-dispersion wavelength is shifted from the 1300-nm region to around 1550 nm. Figure 6-2 shows the effects of dispersion shifting on dispersion. The figure shows that there are two common types of dispersion-shifted fibers: zero-dispersion-shifted fibers and nonzero-dispersion-shifted fibers.

Zero-dispersion-shifted (DS) fibers have the zero dispersion point shifted to 1550 to coincide with the low attenuation operating point. Material dispersion is reduced to zero. DS fibers work well when a single channel—that is, a single data stream—is transmitted through the fiber. Newer systems, however, often send more than one channel through the fiber. They may send separate channels or streams of data at 1546, 1548, 1550, and 1552

FIGURE 6-2

Material dispersion and dispersion-shifted and unshifted fibers

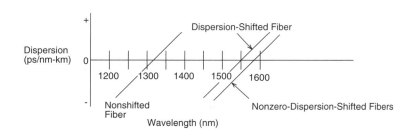

nm. Unfortunately, an effect called four-wave mixing robs the signals of power and increases noise in the systems. Four-wave mixing occurs in fibers that have the zero dispersion point at or near the wavelengths being transmitted.

Figure 6-3 shows the idea of four-wave mixing. When two wavelengths are transmitted through a fiber at the zero-dispersion point, the wavelengths can mix to produce new wavelengths. These wavelengths are a form of noise that can disrupt proper operation of the system. The number of additional wavelengths that can be created is equal to

$$FWM = \frac{N^2(N-1)}{2}$$

where N is the number of original wavelengths. This means that for 2 wavelengths, two additional wavelengths will be produced. But for four wavelengths, 24 new wavelengths can be produced. For eight original wavelengths, the number jumps to 224. Four-wave mixing can seriously limit the use of multiple wavelengths in DWDM applications. In particular, transmission speeds must be lowered.

Adding a small amount of dispersion can suppress four-wave mixing. *Nonzero-dispersion-shifted* (NZ-DS) fibers overcome the problem of four-wave mixing by shifting the zero dispersion point not to 1550 nm, but to a point nearby. For example, fiber can be fabricated to have zero dispersion at 1560 or 1540. Dispersion is significantly reduced to around 1.5 to 3 ps/nm/km and four-wave mixing is not a problem in multiwavelength transmission. NZ-DS fibers, because of their ability to handle high data rates and multiple wavelengths, are widely used in communications applications, surpassing DS fibers.

Figure 6-4 shows the information-carrying capacity as a function of dispersion in a multiwavelength transmission, in this case for four channels. Notice at the zero dispersion point, capacity is severely limited by four-wave mixing. Adding 2 or 3 ps-km of dispersion increases the capacity to allow eight channels, each operating at 2.5 Gbit/s, to be transmitted 1000 km without dispersion compensation.

Dispersion can be compensated for by adjusting the dispersion in positive and negative directions. If the normal fiber has positive dispersion, then adding a compensator with negative dispersion can help reduce the dispersion.

Notice that dispersion varies also with the spectral width of the source. If you limit the range of wavelengths through the fiber, you reduce material dispersion.

FIGURE 6-3
Four-wave
mixing

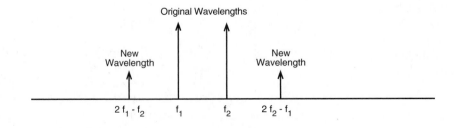

FIGURE 6-4
Effect of four-wave mixing and dispersion shifting on a fiber carrying four channels

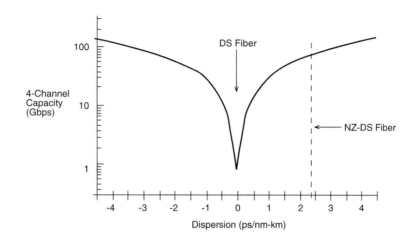

POLARIZATION MODE DISPERSION

This is a minor type of dispersion that only becomes significant in a system that has already minimized other forms of dispersion and that is operating at gigabit data rates. Polarization mode dispersion (PMD) arises from the fact that even a single mode can have *two* polarization states. These polarizations travel at slightly different speeds, thus spreading the signal. A typical PMD value is 0.5 ps/km. In practical terms, for a 100-km transmission distance, PMD limits the signal frequency to 40 GHz.

WAVEGUIDE DISPERSION

Waveguide dispersion is most significant in a single-mode fiber. The energy travels at slightly different velocities in the core and cladding because of the slightly different refractive indices of the materials. Altering the internal structure of the fiber allows waveguide dispersion to be substantially changed, thus changing the specified overall dispersion of the fiber, which is one goal of the advanced single-mode designs discussed in the last chapter.

MODAL NOISE

It would appear that the ideal is always to use a narrow-width source that puts only an extremely narrow beam of a single wavelength into a fiber. With a single-mode fiber, this is generally true: the narrower the range of wavelength, the better. Chromatic dispersion is the main source of dispersion in a single-mode fiber and, therefore, a highly coherent, narrow light source is desirable.

For a multimode fiber, a high-quality laser source is not always desirable. A coherent light coupled into a fiber causes an effect known as speckle patterns. Fiber modes interfere with one another so that the pattern at the end of a fiber is a series of bright and dark spots as shown in Figure 6-5. Speckles themselves are not necessarily bad. What is bad is that the speckle patterns will randomly shift and alter if the output of the laser changes in wavelength (as can happen with temperature changes), if the laser's output pattern changes, or if the fiber vibrates. Thus speckles can change with time. The problem arises at places like connectors, where the loss will vary with changes in the speckle pattern. The amount of power coupled from one fiber to the next will vary as the speckle pattern changes. Thus, modal noise injects random variations in power levels into the system.

Speckling does not occur with incoherent sources like LEDs. Nor does it occur with many of the laser sources available. In one respect, the best lasers—the distributed-feedback lasers used for high-speed, long-distance communications in a single-mode system—are the worst choice for multimode fibers. Thus, high-quality lasers may not always be the best choice.

The speckle pattern is known as modal noise. It represents random variations in the optical power through the fiber.

BANDWIDTH AND DISPERSION

Many fiber and cable manufacturers do not specify dispersion for their multimode offerings. Instead, they specify a figure of merit called the bandwidth-length product or simply bandwidth, given in megahertz-kilometers. A bandwidth of 400 MHz-km means that a 400-MHz signal can be transmitted for 1 km. It also means that the product of the frequency and the length must be 400 or less (BW $\times L \leq 400$). In other words, you can send a lower frequency a longer distance or a higher frequency a shorter distance, as shown in Figure 6-6.

Single-mode fibers, on the other hand, are specified by dispersion. This dispersion is expressed in picoseconds per kilometer per nanometer of source spectral width (ps/km/nm). In other words, for any given single-mode fiber, dispersion is most affected by the source's spectral width: the wider the source width (the greater the range of wavelengths injected into the fiber), the greater the dispersion. Although the conversion of sin-

FIGURE 6-5
Modal noise
from speckle
patterns

FIGURE 6-6
Relation of
bandwidth to
transmission dis-
tance for 400
MHz-km fiber

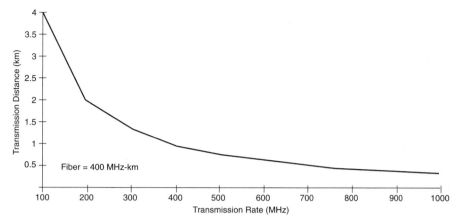

gle-mode dispersion to bandwidth is complex, a rough approximation can be gained from the following equation:

$$BW = \frac{0.187}{(Disp)(SW)(L)}$$

Disp = dispersion at the operating wavelength, in seconds per nanometer per kilometer
SW = the spectral width (rms) of the source, in nanometers
L = fiber length, in kilometers

Here is an example using the following:

Dispersion = 3.5 ps/ns/km
Spectral width = 2 nm
Length = 25 km

Substituting these values into the equation yields

$$BW = \frac{0.187}{(3.5 \times 10^{-12}\text{s/nm/km})(2 \text{ nm})(25 \text{ km})}$$

$$= \quad 1068 \text{ MHz or, roughly, 1 GHz}$$

Doubling the source width to 4 nm significantly reduces the bandwidth to around 535 MHz. So the spectral width of the source has a significant effect on the performance of a single-mode fiber. Reducing the dispersion figure or the source's spectral width increases the bandwidth.

ATTENUATION

Attenuation is the loss of optical power as light travels through the fiber. Measured in decibels per kilometer, it ranges from over 300 dB/km for plastic fibers to around 0.21 dB/km for single-mode fibers.

Attenuation varies with the wavelength of light. Windows are low-loss regions, where fibers carry light with little attenuation. The first generation of optical fibers operated in the first window, around 820 to 850 nm. The second window is the zero-dispersion region of 1300 nm, and the third window is the 1550-nm region. A typical 50/125 graded-index fiber offers attenuation of 4 dB/km at 850 nm and 2.5 dB/km at 1300 nm, which is about a 30% increase in transmission efficiency.

High-loss regions, where attenuation is very high, occur at 730, 950, 1250, and 1380 nm. One wishes to avoid operating in these regions. Evaluating loss in a fiber must be done with respect to the transmitted wavelength. Figure 6-7 shows a typical attenuation curve for a low-loss multimode fiber. Figure 6-8 does the same for a single-mode fiber; notice the high loss in the mode-transition region, where the fiber shifts from multimode to single-mode operation.

Making the best use of the low-loss properties of the fiber requires that the source emit light in the low-loss regions of the fiber.

Plastic fibers are best operated in the visible-light area around 650 nm.

One important feature of attenuation in an optical fiber is that it is constant at all modulation frequencies within the bandwidth. In copper cables, attenuation increases with the frequency of the signal: The higher the frequency, the greater the attenuation. A 25-MHz signal will be attenuated in a copper cable more than will a 10-MHz signal. As a result, signal frequency limits the distance a signal can be sent before a repeater is needed to regenerate the signal. In an optical fiber, both signals will be attenuated the same.

FIGURE 6-7

Attenuation versus wavelength for a multimode fiber (Courtesy of Corning Glass Works)

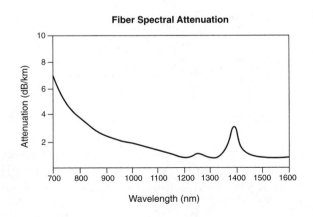

FIGURE 6-8
Attenuation versus wavelength for a single-mode fiber (Courtesy of Corning Glass Works)

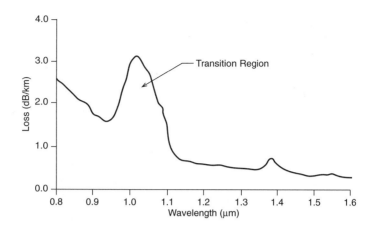

Attenuation in a fiber has two main causes:

1. Scattering
2. Absorption

SCATTERING

Scattering is the loss of optical energy due to imperfections in the fiber and from the basic structure of the fiber. Scattering does just what the term implies: It scatters the light in all directions (Figure 6-9). The light is no longer directional.

Rayleigh scattering is the same phenomenon that causes a red sky at sunset. The shorter blue wavelengths are scattered and absorbed while the longer red wavelengths suffer less scattering and reach our eyes, so we see a red sunset.

Rayleigh scattering comes from density and compositional variations in a fiber that are natural by-products of manufacturing. Ideally, pure glass has a perfect molecular structure and, therefore, uniform density throughout. In real glass, the density of the glass is not perfectly uniform. The result is scattering.

Since scattering is inversely proportional to the fourth power of the wavelength $(1/\lambda^4)$, it decreases rapidly at longer wavelengths. Scattering represents the theoretical lower limits of attenuation, which are as follows:

FIGURE 6-9
Scattering

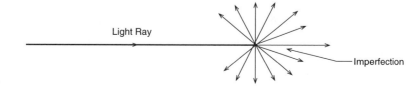

- 2.5 dB at 820 nm
- 0.24 dB at 1300 nm
- 0.012 dB at 1550 nm

ABSORPTION

Absorption is the process by which impurities in the fiber absorb optical energy and dissipate it as a small amount of heat. The light becomes "dimmer." The high-loss regions of a fiber result from water bands, where hydroxyl molecules significantly absorb light. Other impurities causing absorption include ions of iron, copper, cobalt, vanadium, and chromium. To maintain low losses, manufacturers must hold these ions to less than one part per billion. Fortunately, modern manufacturing techniques, including making fibers in very clean environments, permits control of impurities to the point that absorption is not nearly as significant as it was just a few years ago.

MICROBEND LOSS

Microbend loss is that loss resulting from microbends, which are small variations or "bumps" in the core-to-cladding interface. As shown in Figure 6-10, microbends can cause high-order modes to reflect at angles that will not allow further reflection. The light is lost.

Microbends can occur during the manufacture of the fiber, or they can be caused by the cable. Manufacturing and cabling techniques have advanced to minimize microbends and their effects.

EQUILIBRIUM MODE DISTRIBUTION

An important concept about modes in a fiber is that of *equilibrium mode distribution* (EMD). While many modes or paths are available to carry light, not all carry the same amount of energy. Nor do all carry light efficiently. Some modes carry no light—meaning that no energy travels along a potential path. What's more, energy can transfer between modes; it can change paths.

In a perfect fiber, the energy in each mode will stay in that mode. But in a real fiber, energy transfer between modes is caused by bends in the fiber, variations in the diameter or refractive index of the core, or other imperfections.

Over distance, light will transfer between modes until it arrives at EMD. At this point, further transfer of energy between modes does not occur under normal circumstances. It can occur under unusual circumstances such as flaws in the fiber, bends in the cable, and other things. At EMD, inefficient modes have lost their optical energy.

Before it reaches EMD, a fiber is said to be *overfilled* or *underfilled*. An overfilled fiber is one in which marginal modes carry optical energy. This energy will be attenuated or lost over a short distance. You can think of this as excess energy, since for many applications it is unimportant. It will be lost over distance. Some light sources, notably LEDs, can overfill a fiber. This means they inject light into modes that the fiber will not carry efficiently. Some of these modes are in the cladding. Others are high-order modes in the core that will not propagate efficiently.

An underfilled fiber is one in which the light injected into the fiber fills only some of the low-order modes available for propagation of optical energy. A laser, for example, because of its narrow, intense beam, might fill only the lowest-order modes—those traveling with few reflections. Over distance, some of this energy will enter higher-order modes until EMD is reached.

Think of a fiber as a water hose. If you try to couple a flow of water too large for the hose, only some of the water will travel through the hose. Some of the water might travel along the outside of the hose (the cladding), but only for a short distance. If, on the other hand, you shoot a very narrow beam of water into the hose, the water at first does not fill the entire hose. Over distance, however, it will. Thus, the hose will reach a steady state or EMD as the water travels through it.

FIGURE 6-10
Loss and bends

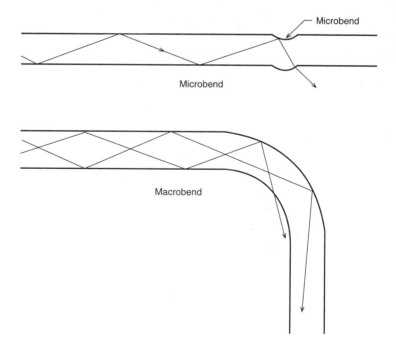

The modal conditions of a short length of fiber depend on the characteristics of the source that injects the light. An LED often overfills the fiber. Over distance, however, the modal conditions become independent of the source.

The distance required to reach EMD varies with the type of fiber. A plastic fiber requires only a few meters or less to reach EMD. A high-quality glass fiber can require tens of kilometers before it reaches EMD.

Equilibrium mode distribution is important to understand for two reasons. First, loss of optical power—attenuation—in an optical fiber depends on modal conditions. In a short length of fiber that has not reached EMD, loss is proportional to length. For a fiber that has reached EMD, loss is proportional to the *square root of length*.

The second reason is the effect that modal conditions has on other conditions in a fiber. Consider the following example as an illustration. Suppose a 1-meter length of fiber is connected to a light source that overfills the fiber. At the end of the fiber we measure the optical energy and find 750 μW of energy. However, this energy includes energy in modes that will be lost by the time the fiber reaches EMD. If, on the other hand, we modify the fiber conditions by wrapping the fiber around a small-diameter mandrel five times, we simulate EMD in this short length of fiber. And now, perhaps, we find only 500 μW of energy emerging from the fiber.

What happened to the other 250 μW? The difference of 250 μW is due to the energy that will be lost before the fiber reaches EMD.

To be able to compare accurately two fibers, two light sources, or two connectors, you must know the conditions under which their manufacturers test them. If one manufacturer uses a fully filled fiber and another uses a fiber under conditions of EMD, the apparent test results will differ dramatically—even if the two fibers are identical. Most fiber-optic test measurements today are performed with fibers at EMD so that comparisons are meaningful.

Another consequence of EMD is the effects it has on two other characteristics of a fiber: numerical aperture and active diameter.

NUMERICAL APERTURE

Numerical aperture (NA) is the "light-gathering ability" of a fiber. Only light injected into the fiber at angles greater than the critical angle will be propagated. The *material* NA relates to the refractive indices of the core and cladding:

$$NA = \sqrt{n_1^2 - n_2^2}$$

Notice that NA is dimensionless.

We can also define the angles at which rays will be propagated by the fiber. These angles form a cone, called the *acceptance cone,* that gives the maximum angle of light acceptance. The acceptance cone is related to the NA:

$$\theta = \text{arcsin (NA)}$$
$$NA = \sin \theta$$

where θ (theta) is the half-angle of acceptance. See Figure 6-11.

The NA of a fiber is important because it gives an indication of how the fiber accepts and propagates light. A fiber with a large NA accepts light well; a fiber with a low NA requires highly directional light.

In general, fibers with a high bandwidth have a lower NA. They thus allow fewer modes. Fewer modes mean less dispersion and, hence, greater bandwidth. NAs range from about 0.50 for plastic fibers to 0.20 for graded-index fibers. A large NA promotes more modal dispersion, since more paths for the rays are provided.

Manufacturers do not normally specify NA for their single-mode fibers (the NA is only about 0.11), because NA is not a critical parameter for the system designer or user. Light in a single-mode fiber is not reflected or refracted, so it does not exit the fiber at angles. Similarly, the fiber does not accept light rays at angles within the NA and propagate them by total internal reflection. As a result, NA, although it can be defined for a single-mode fiber, is not useful as a practical characteristic. Later chapters will show how NA in a multimode fiber is important to system performance and to calculating anticipated performance.

The NA of a fiber changes over distance. High-order modes—those that travel near the critical angle—are often lost. As a graded-index fiber reaches EMD, for example, its NA can be reduced by as much as 50%. This reduction means that the light exiting the fiber does so at angles much less than defined by the acceptance cone. In addition, the spot diameter of light emerging from the fiber can also be reduced. Figure 6-12 shows the diameter of light in the core of a 62.5-μm core under fully filled and EMD conditions. When the fiber is fully filled, the light fills the core. When the fiber reaches EMD, the light diameter is only 50 μm. The NA of the light is similarly reduced.

FIGURE 6-11

Numerical aperture (Illustration courtesy of AMP Incorporated)

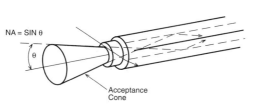
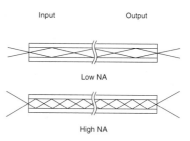

FIGURE 6-12
Active area carrying optical power in a 62.5-μm core under fully filled and EMD conditions (Illustration courtesy of AMP Incorporated)

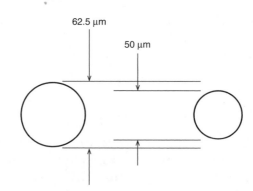

62.5 μm

50 μm

Sources and detectors also have an NA. The NA of a source defines the angles of the exiting light. The NA of the detector defines the angles of light that will operate the detector. Especially for sources, it is important to match the NA of the source to the NA of the fiber so that all the light emitted by the source is coupled into the fiber and propagated. Mismatches in NA are sources of loss when light is coupled from a lower NA to a higher one.

FIBER STRENGTH

One usually thinks of glass as brittle. Certainly, a pane of glass is not easily bent, let alone rolled up. Yet, a fiber can be looped into tight circles without breaking. Furthermore, fiber can be tied into loose knots. (Pulling the knot tight will break the fiber.)

Tensile strength is the ability of a fiber or wire to be stretched or pulled without breaking. The tensile strength of a fiber exceeds that of a steel filament the same size. Furthermore, a copper wire must have twice the diameter to have the same tensile strength as a fiber.

The main causes of weakness in a fiber are microscopic cracks on the surface and flaws within the fiber. Surface cracks are probably most significant. Surface defects can grow with pulling from the tensile load applied as the fiber is installed and from the tensile load on the fiber during its installed lifetime. Temperature changes, mechanical and chemical damage, and normal aging also promote defects.

Defects can grow, eventually causing the fiber to break. If you have ever cut glass, you understand this phenomenon. To cut glass, you make a shallow scribe across the glass. Given a sharp snap, the glass splits along the scribe. The same effect holds true for optical fibers. A flaw acts like the line scribed in the pane of glass. As the fiber is pulled, the flaw grows into the fiber until the fiber breaks.

BEND RADIUS

Even though fibers can be wrapped in circles, they have a minimum bend radius. A sharp bend will snap the glass. Bends have two other effects:

1. They increase attenuation slightly. This effect should be intuitively clear. The bends change the angles of incidence and reflection enough that some high-order modes are lost (similar to microbends).
2. Bends decrease the tensile strength of the fiber. If pull is exerted across a bend, the fiber will fail at a lower tensile strength than if no bend were present.

As a rule of thumb, the minimum bend radius is five times the cable diameter for an *unstressed* cable and 10 times the diameter for a *stressed* cable.

NUCLEAR HARDNESS

Nuclear hardness refers to the ability of equipment to withstand nuclear effects. The effect of nuclear radiation on a conductor is of great interest to the U.S. military (in particular, to protect and maintain its command, control, and communication [C^3] system), to the nuclear power industry, and to other areas of high-radiation risk. Fibers are nonconductive and do not build up static charges when exposed to radiation. Fibers will not short out if their jackets are melted by the high heat of a nuclear emergency.

Fibers do suffer an increase in attenuation when subjected to high levels of continuous nuclear radiation. Nuclear radiation increases the light absorption by the fiber's impurities. The increase in attenuation depends on the amount of the dosage and the intensity of the dosage. A high-intensity burst of 3700 rad in 3 ns increases attenuation to peaks of thousands of decibels per kilometer. This excess attenuation decreases to under 10 dB/km in less than 10 s and to fewer than 5 dB/km in less than 100 s. Thus, glass fibers are able to transmit information after exposure to high-radiation weapons bursts within minutes after the radiation has cleared.

An electromagnetic pulse (EMP) is another concern associated with nuclear weapons, although its effects are more closely related to severe EMI than to nuclear radiation. An EMP results from a nuclear explosion. Two or three nuclear devices detonated over the United States at an altitude of several hundred kilometers could damage or destroy every piece of unprotected electronic equipment in the country.

Gamma rays, produced in the first few nanoseconds after detonation, travel until they collide with electrons of air molecules in the upper atmosphere. These electrons, scattered and accelerated by the collisions, are deflected by the earth's magnetic field to produce a transverse electric current. This current sets up EMPs that radiate downward. Any metal conductors will pick up these pulses and conduct them. A 1-megaton warhead can produce peak fields of 50 kV/m with an instantaneous peak power of around 6 MW/m^2. Such levels are far beyond the ability of equipment to withstand. In short, the entire power and communications grid of the country could be wiped out by an electromagnetic pulse.

SUMMARY

- Dispersion is a general term for phenomena that cause light to spread as it travels through a fiber.
- The four types of dispersion are modal, material, waveguide, and polarization mode.
- Dispersion limits bandwidth.
- Dispersion in multimode fibers includes modal and material dispersion.
- Dispersion in single-mode fibers includes material and waveguide dispersion. Material dispersion is the most important.
- Attenuation is the loss of signal power.
- Attenuation varies with the frequency of light.
- Attenuation does not vary with the signal rate in an optical fiber.
- Numerical aperture is the light-gathering ability of a fiber. It defines the angles at which a fiber will accept and propagate light.
- Fibers have a higher tensile strength than do wires of comparable size.

 Review Questions

1. Name four types of dispersion.
2. Which type of dispersion does not exist in a single-mode fiber?
3. Is the information-carrying capacity of a multimode fiber characterized by dispersion or by bandwidth?
4. Is the information-carrying capacity of a single-mode fiber characterized by dispersion or by bandwidth?
5. If a multimode fiber has a bandwidth of 250 MHz-km, how far can a 750-MHz signal be transmitted?
6. Name the two main mechanisms of attenuation in an optical fiber.
7. Is attenuation lower in a fiber at 850 nm, 1300 nm, or 1550 nm? Why?
8. For any particular single-mode fiber, what specific characteristic limits its data rate?
9. Does an optical fiber recover from a short, intense exposure to nuclear radiation in seconds, minutes, hours, or days?
10. (True/False) The material NA of a fiber always indicates the modal conditions of light traveling within the fiber.

chapter seven

FIBER-OPTIC CABLES

Most often, an optical fiber must be packaged before it is used. Packaging involves cabling the fiber. *Cabling* is an outer protective structure surrounding one or more fibers. It is analogous to the insulation or other materials surrounding a copper wire. Cabling protects copper and fibers environmentally and mechanically from being damaged or degraded in performance. (The additional protection against electrical shock, shorts, and the possibility of fire important to copper is not an issue with dielectric fibers.) The purpose of this chapter is to describe some typical cable structures.

Like their copper counterparts, fiber-optic cables come in a great variety of configurations (Figure 7-1). Important considerations in any cable are tensile strength, ruggedness, durability, flexibility, environmental resistance, temperature extremes, and even appearance.

Evaluation of these considerations depends on the application. An outside plant telephone cable must endure extreme hardship. It must withstand extremes of heat and cold, ice deposits that cause it to sag on a pole, high winds that buffet it, and rodents that chew on it underground. It obviously must be more rugged than a cable connecting equipment within the controlled environment of a telephone switching office. Similarly, a cable running under an office carpet, where people will walk on it and chairs will roll over it, has different requirements than a cable running within the walls of the office.

FIGURE 7-1
Fiber-optic
cables (Photo
courtesy of
Berk-Tek)

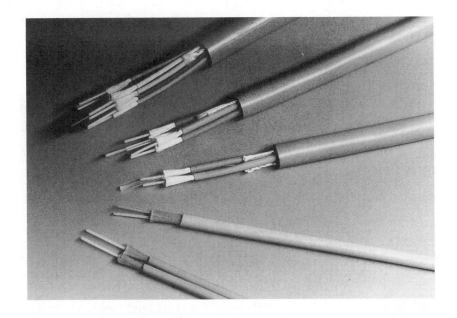

MAIN PARTS OF A FIBER-OPTIC CABLE

Figure 7-2 shows the main parts of a simple single-fiber cable. Even though cables come in many varieties, most have the following parts in common:

- Optical fiber
- Buffer
- Strength member
- Jacket

Since we have already looked extensively at optical fibers, we discuss only the buffer, strength member, and jacket here.

BUFFER

The simplest buffer, discussed in Chapter 5, is the plastic coating applied to the cladding. This buffer, which is part of the fiber, is applied by the fiber manufacturer. An additional buffer is added by the cable manufacturer. (Most cable manufacturers and vendors do not make their own fibers.)

FIGURE 7-2
Parts of a fiber-
optic cable
(Courtesy of
Hewlett-Packard)

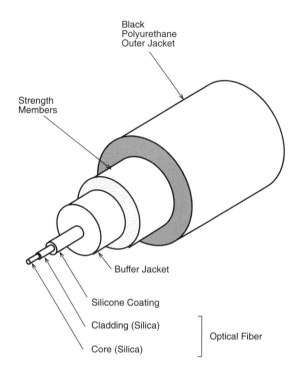

The cable buffer is one of two types: loose buffer or tight buffer. Figure 7-3 shows the two constructions and summarizes the tradeoffs involved with each.

The *loose buffer* uses a hard plastic tube having an inside diameter several times that of the fiber. One or more fibers lie within the buffer tube. The tube isolates the fiber from the rest of the cable and the mechanical forces acting on it. The buffer becomes the load-bearing member. As the cable expands and shrinks with changes in temperature, it does not affect the fiber as much. A fiber has a lower temperature coefficient than most cable elements, meaning that it expands and contracts less. Typically, some excess fiber is in the tube; in other words, the fiber in the tube is slightly longer than the tube itself. Thus the cable can easily expand and contract without stressing the fiber.

The *tight buffer* has a plastic directly applied over the fiber coating. This construction provides better crush and impact resistance. It does not, however, protect the fiber as well from the stresses of temperature variations. Because the plastic expands and contracts at a different rate than the fiber, contractions caused by variations in temperature can result in loss-producing microbends.

FIGURE 7-3
Loose and tight buffers (Courtesy of Belden Electronic Wire and Cable)

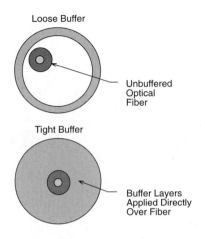

Cable Parameter	Cable Structure	
	Loose Tube	Tight Buffer
Bend Radius	Larger	Smaller
Diameter	Larger	Smaller
Tensile Strength, Installation	Higher	Lower
Impact Resistance	Lower	Higher
Crush Resistance	Lower	Higher
Attenuation Change at Low Temperatures	Lower	Higher

Another advantage to the tight buffer is that it is more flexible and allows tighter turn radii. This advantage can make tight-tube buffers useful for indoor applications where temperature variations are minimal and the ability to make tight turns inside walls is desired.

STRENGTH MEMBERS

Strength members add mechanical strength to the fiber. During and after installation, the strength members handle the tensile stresses applied to the cable so that the fiber is not damaged. The most common strength members are Kevlar aramid yarn, steel, and fiberglass epoxy rods. Kevlar is most commonly used when individual fibers are placed within their own jackets (as in Figure 7-2). Steel and fiberglass members find use in multifiber cables. Steel offers better strength than fiberglass, but it is sometimes undesirable when one wishes to maintain an all-dielectric cable. Steel, for example, attracts lightning, whereas fiberglass does not.

JACKET

The *jacket,* like wire insulation, provides protection from the effects of abrasion, oil, ozone, acids, alkali, solvents, and so forth. The choice of jacket materials depends on the degree of resistance required for different influences and on cost. Figure 7-4 compares the relative properties of various popular jacket materials.

FIGURE 7-4
Properties of jacket materials (Courtesy of Belden Electronic Wire and Cable)

	PVC	LOW DENSITY POLY-ETHYLENE	CELLULAR POLY-ETHYLENE	HIGH DENSITY POLY-ETHYLENE	POLY-PROP-YLENE	POLY-URETHANE	NYLON	TEFLON
Oxidation Resistance	E	E	E	E	E	E	E	O
Heat Resistance	G-E	G	G	E	E	G	E	O
Oil Resistance	F	G	G	G-E	F	E	E	O
Low Temperature Flexibility	P-G	G-E	E	E	P	G	G	O
Weather, Sun Resistance	G-E	E	E	E	E	G	E	O
Ozone Resistance	E	E	E	E	E	E	E	E
Abrasion Resistance	F-G	F-G	F	E	F-G	O	E	E
Electrical Properties	F-G	E	E	E	E	P	P	E
Flame Resistance	E	P	P	P	P	F	P	O
Nuclear Radiation Resistance	G	G	G	G	F	G	F-G	P
Water Resistance	E	E	E	E	E	P-G	P-F	E
Acid Resistance	G-E	G-E	G-E	G-E	E	F	P-F	E
Alkali Resistance	G-E	G-E	G-E	G-E	E	F	E	E
Gasoline, Kerosene, Etc. (Aliphatic Hydrocarbons) Resistance	P	P-F	P-F	P-F	P-F	G	G	E
Benzol, Toluol, Etc. (Aromatic Hydrocarbons) Resistance	P-F	P	P	P	P-F	P	G	E
Degreaser Solvents (Halogenated Hydrocarbons) Resistance	P-F	P	P	P	P	P	G	E
Alcohol Resistance	G-E	E	E	E	E	P	P	E

P = poor   F = fair   G = good   E = excellent   O = outstanding
These ratings are based on average performance of general purpose compounds. Any given property can usually be improved by the use of selective compounding.

When a cable contains several layers of jacketing and protective material, the outer layers are often called the *sheath*. The jacket becomes the layer directly protecting fibers, and the sheath refers to additional layers. This terminology is especially common in the telephone industry.

INDOOR CABLES

Cables for indoor applications include the following:

- Simplex cables
- Duplex cables
- Multifiber cables
- Heavy-, light-, and plenum-duty cables
- Breakout cables
- Ribbon cables

Although these categories overlap, they represent the common ways of referring to fibers. Figure 7-5 shows cross sections of several typical cable types.

SIMPLEX CABLES

Simplex cables contain a single fiber. "Simplex" is a term used in electronics to indicate one-way transmission. Since a fiber carries signals in only one direction, from transmitter to receiver, a simplex cable allows only one-way communication.

DUPLEX CABLES

Duplex cables contain two optical fibers. "Duplex" refers to two-way communication. One fiber carries signals in one direction; the other fiber carries signals in the other direction. (Of course, duplex operation is possible with two simplex cables.) Figure 7-6 compares simplex and duplex operations.

In appearance, duplex cables resemble two simplex cables whose jackets have been bonded together, similar to the jacket of common lamp cords. Ripcord constructions, which allow the two cables to be easily separated, are popular.

Duplex cable is used instead of two simplex cables for aesthetics and convenience. It is easier to handle a single duplex cable, there is less chance of the two channels becoming confused, and the appearance is more pleasing. Remember, the power cord from your lamp is a duplex cable that could easily be two separate wires. Does a single duplex cord in the lamp not make better sense? The same reasoning prevails with fiber-optic cables.

Light Duty Cables

Applications:
- Connections within equipment
- Equipment interconnections within a room
- Connect work stations to wall or floor outlets
- Short distance, intrabuilding cabling
- Interconnections on patch panels and distribution boxes

Single Fiber

Buffer — Strength Members
Optical Fiber — Outer Jacket
0.9
3.0

Dual Fiber

Optical Fiber
Buffer
Strength Members
Outer Jacket
0.9
3.0
6.10

Heavy Duty Cables

Applications:
- Equipment interconnections within a room
- Outdoor use under certain conditions
- Intrabuilding cabling

Single Fiber

Buffer — Strength Members
Optical Fiber — Outer Jacket
0.9
3.7

Dual Fiber

Optical Fiber
Buffer
Strength Members
Outer Jacket
0.9
3.7
7.85

Breakout Cables

Applications:
- Vertical and horizontal fiber distribution; Between floors and wiring centers
- Outdoor use under certain conditions
- Useful where terminations are made directly to equipment or patch panels

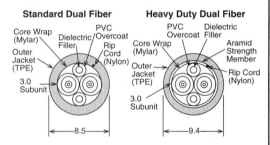

Standard Dual Fiber

Core Wrap (Mylar)
Dielectric Filler
PVC Overcoat
Rip Cord (Nylon)
Outer Jacket (TPE)
3.0 Subunit
8.5

Heavy Duty Dual Fiber

PVC Overcoat
Dielectric Filler
Core Wrap (Mylar)
Aramid Strength Member
Outer Jacket (TPE)
Rip Cord (Nylon)
3.0 Subunit
9.4

Plenum Cables

Applications:
- Use under suspended ceilings and under raised floors
- Intrabuilding cabling between rooms, floors and wiring centers
- Cabling within building air handling spaces

Single Fiber

Buffer — Strength Members
Optical Fiber — Outer Jacket
0.9
3.0

Dual Fiber

Optical Fiber
Buffer
Strength Members
Outer Jacket
0.9
3.0
6.1

DUALAN/ QUADLAN Cables

Applications:
- Offices and equipment rooms
- Fiber-to-desk connections
- Recommended for use with OPTIMATE FSD (Fixed Shroud Duplex) connectors

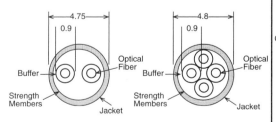

4.75
0.9
Buffer
Optical Fiber
Strength Members
Jacket

4.8
0.9
Buffer
Optical Fiber
Strength Members
Jacket

Plastic Fiber Cables

Applications:
- Optical sensing
- Low cost fiber links
- Internal equipment links
- Industrial controls

Single Fiber

Outer Jacket
Cladding
Core
2.2

Dual Fiber

Outer Jacket
Cladding
Core
2.2
4.6

Note: Dimensions in Millimeters

FIGURE 7-5

Indoor cables (Illustration courtesy of AMP Incorporated)

FIGURE 7-6
Simplex and
duplex

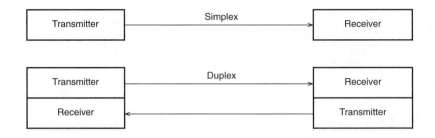

MULTIFIBER CABLES

Multifiber cables contain more than two fibers. They allow signals to be distributed throughout a building. Fibers are usually used in pairs, with each fiber carrying signals in opposite directions. A 10-fiber cable, then, permits five duplex circuits.

Multifiber cables often contain several loose-buffer tubes, each containing one or more fibers. The use of several tubes allows identification of fibers by tube, since both tubes and fibers can be color coded. These tubes are stranded around a central strength member of steel or fiberglass. The stranding provides strain relief for the fibers when the cable is bent.

DUTY SPECIFICATIONS

A cable's construction depends on its application. There are four basic distinctions of application:

- Light duty
- Heavy duty
- Plenum
- Riser

Heavy-duty cables usually have thicker jackets than light-duty cables to allow for rougher handling, especially during installation.

A plenum is the air space between walls, under floor structures, and above drop ceilings. Plenums are popular places to run signal, power, and telephone lines. Unfortunately, plenums are also areas where fires can spread easily through a building.

Certain jacket materials give off noxious fumes when burned. The National Electrical Code (Article 770) requires that cables run through plenums must either be enclosed in fireproof conduits or be insulated and jacketed with low-smoke and fire-retardant materials. Underwriters Laboratories (UL) specifies a procedure called the UL 910 Steiner Tunnel test that rates the flammability of cables. Plenum-rated cables must pass this test and are rated by UL as type OFNP (optical fiber nonconductive plenum) cables.

A riser cable is one that runs vertically between floors of a building. Riser cables must be engineered to prevent fire from spreading between floors and must pass the UL 1666 flame test. They are rated OFNR (optical fiber nonconductive riser) by UL.

BREAKOUT CABLES

Breakout cables have several individual simplex cables inside an outer jacket. The breakout cables shown in Figure 7-5 use two dielectric fillers to keep the cables positioned, while a Mylar wrap surrounds the cables/fillers. The outer jacket includes a ripcord to make its removal fast and easy. The point of the breakout cable is to allow the cable subunits inside to be exposed easily to whatever length is needed. Breakout cables are typically available with two or four fibers, although larger cables also find use.

RIBBON CABLES

Ribbon cable uses a number of fibers side by side in a single jacket. Originally, ribbon cables were used for outdoor cables (see Figure 7-7 later in this chapter). Today, they also find use in premises cabling and computer applications. The cables, typically with up to 12 fibers, offer a very small cross section. They are used to connect equipment within cabinets, in network applications, and for computer data centers. In addition, they are compatible with multifiber array connectors. Ribbon cables are available in both multimode and single-mode versions.

OUTDOOR CABLES

Cables used outside a building must withstand harsher environmental conditions than most indoor cables. Outdoor cables are used in applications such as the following:

- *Overhead:* Cables are strung from telephone poles.
- *Direct burial:* Cables are placed directly in a trench dug in the ground and covered.
- *Indirect burial:* This is similar to direct burial, but the cable is inside a duct or conduit.
- *Submarine:* The cable is underwater, including transoceanic applications.

Such cables must obviously be rugged and durable, since they will be exposed to various extremes. Most cables have additional protective sheaths. For example, a layer of steel armoring protects against rodents that might chew through plastic jackets and into the fiber. Other constructions use a gel compound to fill the cables and eliminate air within the cable. Loose-tube buffers, for example, are so filled to prevent water from seeping into the cable, where it will freeze, expand, and damage the cable. The fibers "float" in the gel that will not freeze and damage the fiber.

Most outdoor cables contain many fibers. The strength member is usually a large steel or fiberglass rod in the center, although small steel strands in the outer sheath are also used. Most high-fiber-count cables divide the fibers among several buffer tubes. Figure 7-7 shows the cross section of several typical constructions.

Another approach is ribbon cable. Here, 12 parallel fibers are sandwiched between double-sided adhesive polyester tape. Each ribbon can be stacked with others to make a rectangular array. For example, a stack of 12 ribbons creates an array of 144 fibers. This array is placed in a loose tube, which in turn is covered by two layers of polyethylene. Each polyethylene layer contains 14 steel wires serving as strength members. Depending on the application, additional layers, such as steel armoring, cover the polyethylene. The first commercial fiber-optic system installed by AT&T, in 1977, used a ribbon structure. Figure 7-8 shows such ribbon cable.

In cables containing many fibers, not all the fibers are always used at first. Some are kept as spares to replace fibers that fail in the future. Others are saved for future expansion, as the demands for additional capacity dictate. Installing a cable can be expensive. Having extra fibers in place when they are needed saves future installation costs of additional fibers and cables. It pays to think ahead.

ADDITIONAL CABLE CHARACTERISTICS

Lengths

Cables come reeled in various lengths, typically 1 or 2 km, although lengths of 5 or 6 km are available for single-mode fibers. Long lengths are desirable for long-distance applications, since cables must be spliced end to end over the run. Each splice introduces additional loss into the system. Long cable lengths mean fewer splices and less loss.

Color Coding

Fiber coatings or buffer tubes or both are often color coded to make identification of each fiber easier. In a long-distance link, one must ensure that fiber A in the first cable is spliced to fiber A in the second cable, B to B, C to C, and so forth. Color coding simplifies fiber identification.

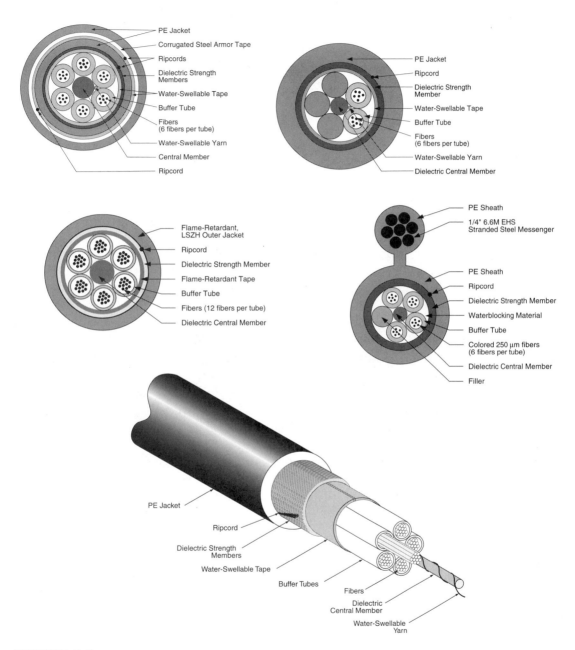

FIGURE 7-7
Outdoor multifiber cables (Courtesy of Siecor Corporation)

FIGURE 7-8
Ribbon cable
(Courtesy of
AT&T/Lucent
Technologies
Bell Labs)

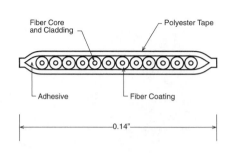

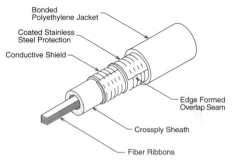

Loads

Most cable vendors specify the maximum tensile loads that can be applied to the cable. Two loads are usually specified. The *installation load* is the short-term load that the fiber can withstand during the actual process of installation. This load includes the additional load that is exerted by pulling the fiber through ducts or conduits, around corners, and so forth. The maximum specified installation load limits the length of cable that can be installed at one time, given the particular application. Different applications will offer different installation load conditions. One must carefully plan the installation to avoid overstressing the fiber.

The second load specified is the long-term or *operating load*. During its installed life, the cable cannot withstand loads as heavy as it withstood during installation. The specified operating load is therefore less than the installation load. The operating load is also called the *static load*.

Installation and operational loads are specified in pounds or newtons. Allowable loads, of course, depend on cable construction and intended application. A typical specification for a simplex indoor cable is an installation load of 250 lb (1112 N) and an operational load of 10 lb (44 N).

HYBRID CABLES

Fiber-optic cables sometimes also contain copper conductors, such as twisted pairs. Although the conductors can be used for routine communications, they have two other popular uses. One use is to allow installers to communicate with each other during installation of the fiber, especially with long-distance telephone installation. A technician performing a splice in a manhole must often be in communication with a switching office several miles away. The office contains the test equipment, operated by another technician, that tests the quality of the splice. The other use is to power remote electronic equipment such as repeaters.

UNDERSTANDING CABLE SPECIFICATIONS

Most cable configurations come with various sizes and types of fibers. For example, many fibers have a buffer coating of 250- or 900-μm diameter. This coating allows fibers of 8/125, 50/125, 62.5/125, or 100/140 μm to be used. Each of these fibers can further be offered with various attenuations and bandwidths to satisfy the needs of a particular application. In addition, a cable using a loose-tube buffer can hold one or several fibers. None of these factors significantly influences cable construction. The same construction can accommodate all these differences easily.

As fiber-optic technology became widespread, serious debate evolved over which multimode fiber was best suited to different applications. For example, 62.5/125- and 100/140-μm fibers were all proposed for premises wiring and local area networks. The debate centered on the technical and costs merits of each fiber: attenuation, bandwidth, NA, ease and cost of coupling light into the fiber, and so forth.

The "winner" of these debates was the 62.5/125-μm fiber, which is the specified or preferred fiber in nearly all applications involving premises wiring, LANs, computer inter-connections, and similar uses. 50/125 fiber is making a "comeback" because of its higher bandwidth.

Single-mode fibers are still the preferred choice for long-distance, high-speed applications, while both 50/125- and 100/140-μm are used in many applications.

Fiber-optic cables are typically offered with standard-grade and premium-grade fibers. Many application standards specify performance that is met by standard-grade cables. In most cases, the cable performance is a minimum, the cable may exceed the stated performance for a requirement such as bandwidth. For example, most standards call for a 62.5-μm cable to have a minimum bandwidth of 160 MHz at 850 nm and 300 MHz at 1300 nm. The standard for FDDI networks calls for a 500-MHz bandwidth at 1300 nm. It is possible to buy cable with a bandwidth of 200 MHz at 850 mn and 600 MHz at 1300 nm. Similarly, 50/125-μm is available with a standard bandwidth of 500 MHz or an extended bandwidth of 1000 MHz (at both 850 and 1300 nm). The point is that it is possible to buy cabled fiber at different levels of performance for the same type of fiber.

SUMMARY

- Cabling is the packaging of optical fibers for their intended application.
- The main parts of the cable are the buffer, strength member(s), and jacket.
- Cables are available for most application environments.

Review Questions

1. Name three materials commonly used as strength members.
2. Name two types of cable buffers.
3. Which is greater: installation load or operating load?
4. A main difference between an indoor cable and an outdoor cable is
    A. Fiber bandwidth
    B. Ruggedness and durability
    C. Number of fibers
    D. Attenuation
5. Describe the difference between simplex and duplex cables.
6. Name two uses for copper wires used in a fiber-optic cable.

chapter eight

SOURCES

At each end of a fiber-optic link is a transducer, which is simply a device for converting energy from one form to another. The source is an electro-optic transducer—that is, it converts the electrical signal to an optical signal. The detector at the other end is the optoelectronic transducer. It converts optical energy to electrical energy.

The source is either a light-emitting diode (LED) or a laser diode. Both are small semiconductor chips, about the size of a grain of sand, that emit light when current is passed through them. To help you understand the operation of LEDs and lasers, as well as the photodetectors described in Chapter 9, we must first review some of the fundamentals of materials in general and semiconductors in particular.

SOME ATOMIC MATTERS

An atom consists of an inner nucleus around which swirls a cloud of electrons. The electrons circle the nucleus in shells. As shown in Figure 8-1, each shell has a maximum number of electrons in it. The inner shell, K, has a maximum of 2 electrons, L a maximum of 8, M a maximum of 18, and N a maximum of 32. More important for our purposes here, however, is that the outer shell always has a maximum of 8 electrons. This outer shell is called the *valence shell* or *valence band*. The valence band is involved in the chemical bonds that allow elements and compounds to exist.

For an electron to flow as current, it must be free from the valence band to be able to move about in the molecular structure of the material. An electron so freed from the atom is said to be *in the conduction band*. The electron is a free electron because it no longer is bound to the atom. To be available, then, the electron must move from the valence band to

FIGURE 8-1

Electronic struc-
ture of an atom

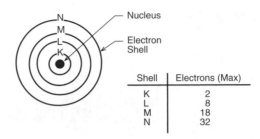

Shell	Electrons (Max)
K	2
L	8
M	18
N	32

the conduction band. How easily this occurs is what separates good conductors from poor conductors.

An atom with only one electron in the valence shell is a very good conductor. Only a small amount of energy is required to promote the electron to the conduction band as a free electron. Copper, gold, and silver, having only one valence electron, are good conductors.

Iron, cobalt, and platinum are examples of elements with eight valence electrons. As a result, they are poor conductors, because an appreciable amount of external energy is required to separate a bound electron. An atom "wishes" to have a filled outer shell. For copper, it easily releases the outer electron, so the next inner shell is filled. For iron, its shell is filled, and it does not wish to release an electron.

A *semiconductor* is a material whose properties fall between those of good conductors like copper and poor conductors like plastics. Separating an electron from the valence shell requires a medium amount of energy. Silicon, for example, has four electrons in its outer shell.

Energy is involved in the movement of an electron between the valence and conduction bands. For an electron to move from the valence band to the conduction band, external energy must be added. An atom that has lost a valence electron develops forces to attract a free electron and complete its shell. When an electron moves from the conduction band to the valence band, it emits energy. The energy released depends on the difference in energy levels between the two bands involved.

Figure 8-2 shows the two energy bands or levels of interest for our understanding of sources and detectors. For an electron to move from one band to another, energy must be either emitted or absorbed. The energy required is the difference between the energy levels of the two bands: $E_1 - E_2$. This difference is the *band gap*. If the band gap is proper—that is, if the difference between the band gap is in a given range—the energy is optical energy, or light. The wavelength of the energy is equal to

$$\lambda = \frac{hc}{|E_1 - E_2|}$$

FIGURE 8-2
Absorption and
emission

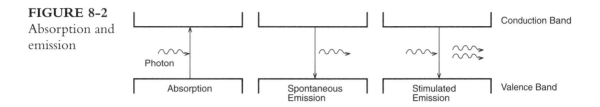

where h is Planck's constant (6.625×10^{-34} J-s), c is the velocity of light, and E_1 and E_2 are the energy levels.

In the spontaneous emission shown in Figure 8-2, an electron randomly moves to the lower energy level. An LED works by spontaneous emission. In stimulated emission, a photon of light causes, or stimulates, an electron to change energy levels. A laser operates on the principle of stimulated emission.

In absorption, external energy, such as supplied by a photon of incident light, supplies the energy to allow the electron to move to the higher energy band.

SEMICONDUCTOR PN JUNCTION

The semiconductor pn junction is the basic structure used in the electro-optic devices for fiber optics. Lasers, LEDs, and photodiodes all use the pn junction, as do other semiconductor devices such as diodes and transistors. We will first describe the basic operation of the junction, and then we will describe the operation of LEDs and lasers. Photodiodes are discussed in greater detail in Chapter 9.

As we saw earlier, a silicon atom has four valence electrons. It is these electrons that form the bonds that hold the atoms together in the crystalline structure of the element. For silicon, these bonds are covalent bonds in which the atoms share their electrons in the bond, as shown in Figure 8-3. Here each atom has access to eight valence electrons required for a full shell, four of its own and four from surrounding atoms. All the electrons are taken up in covalent bonds, and none are readily available as free electrons.

The pn junction begins with such a material as silicon. Suppose we add a material with five valence electrons to a silicon crystal. Since only four electrons are needed for the covalent bonds, an electron is left. This electron is a free electron that can move around the structure. It is really a free electron in the conduction band. Because this material has an excess of negatively charged electrons, it is called *n-type* material.

Suppose, on the other hand, we add a material with only three valence electrons to the silicon. One of the covalent bond sites will not be filled by an electron because none is available. This vacancy is called a *hole*. A hole in a semiconductor is a very strange thing because, by definition, it is an absence of something. In appearance, however, it is a charge

FIGURE 8-3
Covalent bonds
in a silicon atom

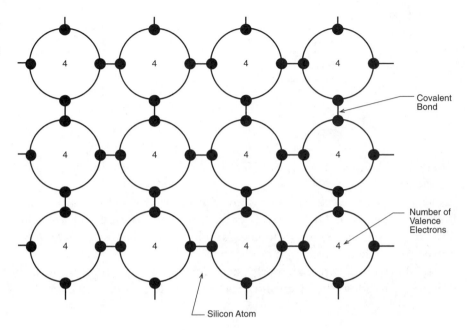

carrier similar to an electron, but it is positively charged. It appears as a positively charged particle. The material appears to have an excess of positively charged holes, so it is called *p-type* material.

The pn junction of a diode is formed when a semiconductor material is purposely doped with atoms to form p-type and n-type areas separated by a junction. The n-type material has free electrons, and the p-type material has holes. When the p-type and n-type materials are brought together, holes and electrons sweep across the junction and recombine. Recombination means that a free electron "falls" into a hole, moving from the conduction band to the valence band. It becomes part of the covalent bonding structure of the atom. Both the hole and the electron disappear as charge carriers.

Because of the recombination of carriers around the pn junction, no carriers exist in the area. A barrier is formed within this depleted region that prevents further migration of electron and hole carriers across the junction unless additional energy is applied to the material.

When a hole and electron recombine, energy is emitted. Depending on the material, this energy may or may not be light. For silicon, the energy is not light. It is heat in the form of vibrations in the crystal structure.

Lasers and LEDs use elements from Groups III and V of the periodic table. These elements have three and five electrons in their valence shells. If we combine an equal number of atoms with three electrons and an equal number of atoms with five electrons, we will have a situation similar to that of silicon. The atoms will form covalent bonds so that the atoms have filled valence shells. No free carriers exist. To create an n-type material, we combine Group V materials in greater proportion to Group III atoms. Now the structure has free electrons available as carriers. Similarly, a structure having more Group III atoms than Group V atoms will result in holes being available as carriers. Figure 8-4 shows these three combinations using gallium arsenide. Gallium atoms have three valence electrons; arsenic atoms have five valence electrons.

Gallium arsenide has a structure and band gap between the conduction and valence bands such that the recombination of carriers results in emission of light.

An LED is a pn semiconductor that emits light when a *forward bias* is applied (the negative battery terminal connects to the n-type material). In a forward bias, electrons are injected into the n-type material and extracted from the p-type material. (Extracting elec-

FIGURE 8-4
Holes and free electrons in GaAs

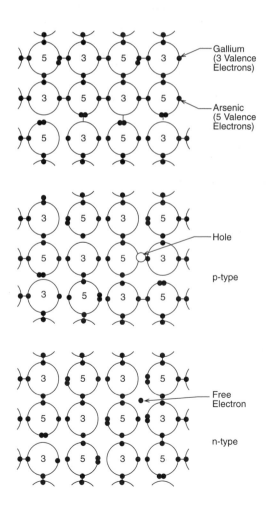

trons from p-type material is the same as injecting holes into the material.) Figure 8-5 shows the schematic symbol for an LED and its bias arrangement.

The forward bias causes the electrons and holes to move toward one another and cross the *depletion area* of the junction. Electrons and holes combine, emitting light in the process. For the operation to be maintained, current must continually be supplied to maintain an excess of carriers for recombination. When the drive is removed, recombination of carriers around the junction recreates the depletion area and emission ceases.

LEDS

LEDs used in fiber optics are somewhat more complex than the simple device we have just described, although their operation is essentially the same. The complexities arise from the desire to construct a source having characteristics compatible with the needs of a fiber-optic system. Principal among these characteristics are the wavelength and pattern of the emission.

The LED we have described is a *homojunction* device, meaning that the pn junction is formed from a single semiconductor material. Homojunction LEDs are surface emitters giving off light from the edges of the junction as well as its entire planar surface. The result is a low-radiance output whose large pattern is not well suited for use with optical fibers. The problem is that only a very small portion of the light emitted can be coupled into the fiber core.

A *heterojunction* structure solves this problem by confining the carriers to the active area of the chip. A heterojunction is a pn junction formed with materials having similar crystalline structure but different energy levels and refractive indices. The differences confine the carriers and provide a more directional output for the light. The difference in refractive index, for example, can be used to confine and to guide the light in much the same manner that it is confined and guided in an optical fiber. The result is a high-radiance output.

Figure 8-6 shows both a surface-emitting and an edge-emitting LED. The edge-emitting diode uses an active area having a stripe geometry. Because the layers above and

FIGURE 8-5
Light-emitting
diode

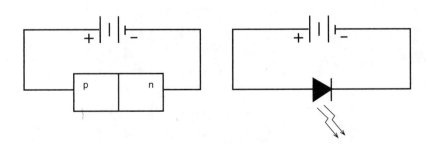

FIGURE 8-6
LEDs and lasers

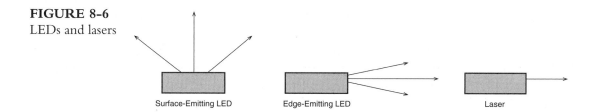

Surface-Emitting LED   Edge-Emitting LED   Laser

below the stripe have different refractive indices, carriers are confined by the waveguide effect produced. (The waveguide effect is the same phenomenon that confines and guides the light in the core of an optical fiber.) The width of the emitting area is controlled by etching an opening in the silicon oxide insulating area and depositing metal in the opening. Current through the active area is restricted to the area below the metal film. The result is a high-radiance elliptical output.

The materials used in the LED influence the wavelengths emitted. LEDs emitting in the first window of 820 to 850 nm are usually gallium aluminum arsenide (GaAlAs). "Window" is a term referring to ranges of wavelengths matched to the properties of the optical fiber. One reason this region was the first exploited in fiber optics is that these devices were better understood, more easily manufactured, more reliable, and less costly. In short, the 820-nm technology was mature, as the 1300-nm technology now is. The movement from 1300 to 1550 nm depends in part on improved technology for sources.

Long-wavelength devices for use at 1300 nm are made of gallium indium arsenide phosphate (GaInAsP) and other combinations of Group III and Group V materials.

LASERS

Laser is an acronym for *l*ight *a*mplification by the *s*timulated *e*mission of *r*adiation. Lasers (Figure 8-6) provide stimulated emission rather than the simpler spontaneous emission of LEDs. The main difference between an LED and a laser is that the laser has an optical cavity required for lasing. This cavity is formed by cleaving the opposite end of the chip to form highly parallel, reflective mirrorlike finishes.

At low drive currents, the laser acts like an LED and emits light spontaneously. As the current increases, it reaches the threshold level, above which lasing action begins. A laser relies on high current density (many electrons in the small active area of the chip) to provide lasing action. Some of the photons emitted by the spontaneous action are trapped in the Fabry-Perot cavity, reflecting back and forth from end mirror to end mirror. These photons have an energy level equal to the band gap of the laser materials. If one of these photons influences an excited electron, the electron immediately recombines and gives off a photon. Remember that the wavelength of a photon is a measure of its energy. Since the energy of the stimulated photon is equal to the original, stimulating photon, its wavelength is equal to that of the original, stimulating photon. The photon created is a *duplicate of the first photon: It has the same wavelength, phase, and direction of travel.* In other words, the incident photon has stimulated the emission of another photon. Amplification has occurred, and emitted photons have stimulated further emission.

The high drive current in the chip creates population inversion. *Population inversion* is the state in which a high percentage of the atoms moves from the ground state to the excited state so that a great number of free electrons and holes exists in the active area around the junction. When population inversion is present, a photon is more likely to stimulate emission than be absorbed. Only above the threshold current does population inversion exist at a level sufficient to allow lasing.

Although some of the photons remain trapped in the cavity, reflecting back and forth and stimulating further emissions, others escape through the two cleaved end faces in an intense beam of light. Since light is coupled into the fiber only from the "front" face, the "rear" face is often coated with a reflective material to reduce the amount of light emitted. Light from the rear face can also be used to monitor the output from the front face. Such monitoring can be used to adjust the drive current to maintain constant power level on the output.

Thus, the laser differs from an LED in that laser light has the following attributes:

- *Nearly monochromatic:* The light emitted has a narrow band of wavelengths. It is nearly *monochromatic*—that is, of a single wavelength. In contrast to the LED, laser light is not continuous across the band of its spectral width. Several distinct wavelengths are emitted on either side of the central wavelength.
- *Coherent:* The light wavelengths are in phase, rising and falling through the sine-wave cycle at the same time.
- *Highly directional:* The light is emitted in a highly directional pattern with little divergence. *Divergence* is the spreading of a light beam as it travels from its source.

SOURCE CHARACTERISTICS

This section describes some of the main characteristics of interest for sources. In doing so, it compares LEDs and lasers. These characteristics help determine the suitability of an LED or a laser for a given application. Figure 8-7 provides a brief comparison of laser and LED characteristics.

FIGURE 8-7
Relative charac-
teristics: LED
and laser

CHARACTERISTIC	LED	LASER
Output power	Lower	Higher
Speed	Slower	Faster
Output pattern (NA)	Higher	Lower
Spectral width	Wider	Narrower
Single-mode compatibility	No	Yes
Ease of Use	Easier	Harder
Lifetime	Longer	Long
Cost	Lower	Higher

OUTPUT POWER

Output power is the optical power emitted at a specified drive current. As shown in Figure 8-8, an LED emits more power than a laser operating below the threshold. Above the lasing threshold, the laser's power increases dramatically with increases with drive current. In general, the output power of the device is in the following decreasing order: laser, edge-emitting LED, surface-emitting LED.

FIGURE 8-8
Power output
versus drive cur-
rent (Illustration
courtesy
of AMP
Incorporated)

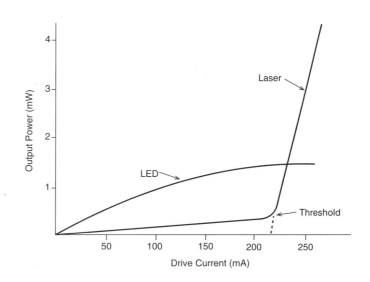

OUTPUT PATTERN

The output pattern of the light is important in fiber optics. As light leaves the chip, it spreads out. Only a portion actually couples into the fiber. A smaller output pattern allows more light to be coupled into the fiber. A good source should have a small emission diameter and a small NA. The emission diameter defines how large the area of emitted light is. The NA defines at what angles the light is spreading out. If either the emitting diameter or the NA of the source is larger than those of the receiving fiber, some of the optical power will be lost. Figure 8-9 shows typical emission patterns for an LED and a laser.

The loss of optical power from mismatches in NA and diameter between the source and the core of multimode fiber is as follows: When the diameter of the source is greater than the core diameter of the fiber, the mismatch loss is

$$\text{loss}_{\text{dia}} = 10 \log_{10} \left(\frac{\text{dia}_{\text{fiber}}}{\text{dia}_{\text{source}}} \right)^2$$

No loss occurs when the core diameter of the fiber is larger.

When the NA of the source is larger than the NA of the fiber, the mismatch loss is

$$\text{loss}_{\text{NA}} = 10 \log_{10} \left(\frac{\text{NA}_{\text{fiber}}}{\text{NA}_{\text{source}}} \right)^2$$

No loss occurs when the fiber NA is the larger.

Consider, for example, a source with an output diameter of 100 μm and an NA of 0.30 that is connected to a fiber with a core diameter of 62.5 μm and an NA of 0.275. The losses from diameter and NA mismatches are as follows:

$$
\begin{aligned}
\text{loss}_{\text{dia}} &= 10 \log_{10} \left(\frac{62.5}{100} \right)^2 \\
&= 10 \log_{10} (0.390625) \\
&= -4.1 \text{ dB} \\
\text{loss}_{\text{NA}} &= 10 \log_{10} \left(\frac{0.275}{0.30} \right)^2 \\
&= 10 \log_{10} (0.934444) \\
&= -0.8 \text{ dB}
\end{aligned}
$$

The total mismatch loss is 4.9 dB. If the source output power is 800 μW, only about 260 μW are coupled into the fiber.

A single-mode fiber requires a laser source. The laser provides a small, intense beam of light compatible with the small core of the fiber. The output of most lasers is elliptical, rather than circular. The vertical-cavity surface-emitting laser emits a circular pattern.

FIGURE 8-9
Emission patterns

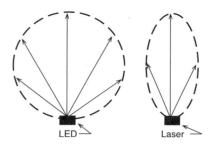

SPECTRAL WIDTH

As the discussion of material dispersion in Chapter 6 emphasized, different wavelengths travel through a fiber at different velocities. The dispersion resulting from different velocities of different wavelengths limits bandwidth. Lasers and most LEDs do not emit a single wavelength; they emit a range of wavelengths. This range is known as the *spectral width* of the source. It is measured at 50% of the maximum amplitude of the peak wavelength. For example, if a source has a peak wavelength of 850 nm and a spectral width of 30 nm, its output ranges from 835 to 865 nm.

At 850 nm, material dispersion can be estimated as 0.1 ns/km for each nanometer of source spectral width. An LED with a 30-nm spectral width results in 3 ns of dispersion over a 1-km run.

Figure 8-10 shows that the spectral width of a laser is substantially narrower than that of an LED. The spectral width of a laser is 0.1 to 5 nm, but that of an LED is tens of nanometers. As a rule of thumb, spectral width is not important in fiber-optic links running under 100 MHz for only a few kilometers. Spectral width is especially important in high-speed, long-distance, single-mode systems because the resulting dispersion is the principal limiting factor on system speed. Remember that the practical bandwidth of a single-mode fiber is specified as dispersion in picoseconds per kilometer per *nanometer of source width* (ns/km/nm).

Because laser spectral width is a main limiting characteristic of long-distance, high-speed, single-mode systems, great effort in recent years has been expended to devise and to build reliable single-wavelength laser diodes. Such devices use advanced structures to promote the center wavelength and to suppress others.

SPEED

The source must turn on and off fast enough to meet the bandwidth requirements of the system. Source speed is specified by rise and fall times. Lasers have rise times of less

FIGURE 8-10
Typical spectral widths (Illustration courtesy of AMP Incorporated)

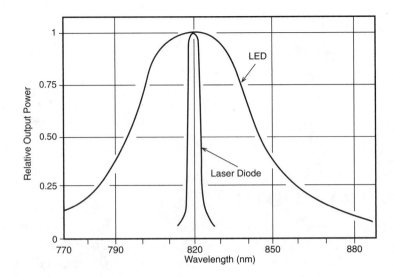

than 1 ns, whereas slower LEDs have rise times of nanoseconds. A rough approximation of bandwidth for a given rise time is

$$BW = \frac{0.35}{t_r}$$

where a rise time in nanoseconds gives a bandwidth in gigahertz. For example, a 1-ns rise time allows 350-MHz operation, and a 5-ns rise time allows 70-MHz operation.

LIFETIME

The expected operating lifetime of a source runs into the millions of hours. Over time, however, the output power decreases because of increasing defects in the device's crystalline structure. The lifetime of the source is typically taken to end when the peak output power is reduced 50% or 3 dB. An LED emitting a peak power of 1 mW, for example, is considered at the end of its lifetime when its peak power becomes 500 μW.

EASE OF USE

Although a laser provides better optical performance than an LED, it is also more expensive, less reliable, and harder to use. Lasers have a shorter expected lifetime than LEDs. They also require more complex transmitter circuits. For example, the output power of a laser can change significantly with temperature. Maintaining proper output levels over the required temperature range requires circuitry that detects changes in output and adjusts the drive current accordingly. One method of maintaining these levels is with a photodiode to monitor the light output on the back facet of the laser. The current from the photodiode changes with variations in light output. These variations provide feedback to adjust the laser drive current.

TYPES OF LASER DIODES

While many types of semiconductor lasers have been created, five deserve special mention because of their wide use in fiber optics.

FABRY-PEROT LASERS

The Fabry-Perot laser is the most common type. The name derives from the way the laser cavity is formed. The Fabry-Perot laser emits more than a single wavelength. It emits a few, less intense, wavelengths on either side of the center wavelength to achieve a spectral width on the order of 3 to 6 nm. For example, if the center wavelength is 1310 nm, other emitted wavelengths might be 1307, 1309, 1311, and 1313.

You might consider the Fabry-Perot laser to be a general-purpose laser for fiber optics. It has a narrow spectral width, it is moderately fast and able to achieve speeds in the gigabit range, and it is moderately priced. The spectral width, however, causes material (chromatic) dispersion. This dispersion limits the bandwidth of the fiber.

One phrase you'll sometimes see describing a laser's structure is *multiquantum well* or *MQW.* A quantum well is a way of building a structure to contain and guide the emitted photons. Multiple quantum wells simply offer a more complex structure for achieving a higher degree of precision and efficiency.

DISTRIBUTED FEEDBACK LASERS

The distributed feedback (DFB) laser occupies the high end, both in terms of performance and price. If the Fabry-Perot laser is a high-end family car, the DFB laser is a sleek sports car. It emits a single wavelength with a spectral width under 1 nm. The lasing cavity uses an internal grating structure so that only one wavelength is amplified while all others are suppressed. The wavelength emitted depends on the length of the cavity and the spacing of the grating. Figure 8-11 shows the structure of a DFB laser, with its internal grating.

In addition, DFB lasers have a high output, capable of coupling several mW of power into a single-mode fiber. Because of their high speed, narrow spectral width, and high output powers, DFB lasers are the favored type for telecommunications and CATV applica-

FIGURE 8-11
DFB laser

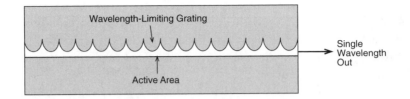

tions. Digital DFB lasers typically have an output of 1 to 5 mW, while analog versions can have outputs over 10 mW.

DFB lasers usually have the fiber included as a pigtail to the package. During assembly, the fiber is precisely aligned with the output of the laser chip. This type of alignment—called active alignment—requires the laser be operated as the fiber is positioned so that the output can be monitored to achieve the most efficient coupling of optical power from the laser into the fiber. Active alignment is labor intensive, even with a degree of automation.

DFB lasers are available for both 1300 and 1550 transmission. In the 1550-nm region, in particular, they are available with precisely defined wavelengths. As you will see in Chapter 16, this is an important feature for increasing the capacity of a fiber-optic system. It is possible to transmit multiple wavelengths, each separate from the other. In other words, you can transmit a 2.5-Gbps signal at 1550 nm. You can transmit a second at 1552 nm, a third at 1554 nm, and a fourth at 1556 nm. This yields a data rate of 10 Gbps. But doing so obviously requires lasers with precise and tightly controlled output, which is met by the DFB laser. One manufacturer offers one model of a DFB laser with a choice of over 38 output wavelengths corresponding to the ITU channel wavelengths. These range from 1532.68 nm to 1562.23 nm, with about 0.8 nm difference between one laser and the next. (We'll discuss the importance of these ITU wavelengths in Chapter 16. For now, note that DFB lasers have precise and narrow spectral outputs.)

DFB lasers are expensive, easily running several thousand dollars for a single package.

CD LASERS

These are the lasers used in compact disk players for music and computer CD-ROMs. They typically emit a narrow spectrum at 790 nm, although they are also available at 850 nm. Since 790 nm is not an attractive window for fiber optics, what is the attraction of the CD laser? In a word: cost. CD lasers are mass produced by the millions, which significantly lowers their price. They are fast. For applications like LANs, where required transmission distances rarely exceed a couple of kilometers in multimode applications, the CD laser works well.

While we typically associate lasers with single-mode fibers, CD lasers are used with multimode fibers. Both 50-μm- and 62.5-μm-core fibers are used. The 790-nm wavelength is below the cutoff frequency of a single-mode fiber so that its operation will be multimode in any event.

VERTICAL-CAVITY SURFACE-EMITTING LASERS

Vertical-cavity surface-emitting lasers are usually called a VCSEL ("vick-sil"). Its name implies the essential difference between it and other semiconductor lasers. A typical laser has a horizontal cavity and emits from the edge (but isn't called a HCEEL). The lasing cavity of the VCSEL runs from top to bottom. The bottom is mirrored so that the main emission is from the top of the device—hence the name.

The VCSEL is one of the newer types of semiconductor lasers, only becoming commercially available in the mid- to late 1990s. But it has several characteristics that make it attractive as a low-cost, high-performance device.

When a semiconductor laser is manufactured, it is "grown" on a circular disk called a wafer. A single wafer can contain hundreds or thousands of individual lasers. Traditional edge-emitting lasers cannot be tested until the wafer is sliced into the individual lasers. Yield—the ratio of good lasers to the total number manufactured—is low. Therefore, a great deal of effort is required to even begin separating the good lasers from the rejects. VCSELs can be tested at the wafer stage so that only good ones need to be separated. This lowers costs at the initial manufacturing stage.

In addition, VCSELs can be applied in arrays. Several VCSELs, rather than being separated, can be used in a data link together without being separated at the wafer stage. The array can be one- or two-dimensional—one single row or several rows. While most transmitter packages have a single LED or laser, a VCSEL-based package can have up to 12 lasers.

The ability to use VCSELs in arrays opens up some new possibilities. One is parallel transmission, sending individual bits of a byte on different fibers. Most systems today operate serially, sending all bits sequentially over a single fiber. Parallel transmission allows an increase in data rates. If you send 800 Mbit/s over a serial link, you can also say your rate is 100 M*bytes*/s since it takes 8 bits for each byte. If you can send the data in parallel and still maintain 800 Mbit/s over each fiber, your data rate is now 800 Mbyte/s since you are using eight fibers. Each bit of a byte is sent at the same time over a different fiber. The total bit rate is 6400 Mbit/s.

You can also treat each laser in the transmitter as a separate serial device. Thus an 8-laser transmitter can operate in a byte-oriented parallel fashion or as eight separate serial lines.

The ability to package multiple lasers in a single package means a lower cost in application. Rather than having to buy eight transmitters and receivers, you can buy a single package of each. What's more, since the eight VCSELs come from the same wafer, they have the same characteristics, with very little variation. This can be important since you want the signal into each fiber to be the same.

The first crop of VCSELs operate in the 850-nm window and are intended for use with multimode fibers. Future generations will probably operate at 1300 nm and other wavelengths, on both single-mode and multimode fibers. One additional benefit of a VCSEL over an edge-emitting laser is that its output is round rather than elliptical. The round output means that most of the output energy is easily coupled into a 50- or 62.5-μm core. While the output beam of the VCSEL is wider than that of a Fabry-Perot or DFB laser, it still offers exceptional coupling efficiency. Thanks to the VCSEL, 50/125-μm fibers have found renewed interest for higher bandwidth LAN applications. The 62.5/125-μm fiber was preferred because it allowed more light from an LED to be coupled into the fiber. The drawback was a low bandwidth at the shorter 850-wavelength. VCSELs offer superior coupling into a 50/125-μm fiber. At 850-nm, a 62.5/125 fiber has a bandwidth of 160 MHz-km; the 50/125 fiber has a bandwidth of 500 MHz-km. Figure 8–12 compares the output of a VCSEL and a typical LED.

You can think of a VCSEL as bridging the gap between LEDs and lasers. They have the lower costs and compatibility with multimode fibers that you associate with LEDs. They have the speed, narrow spectral width, and narrow output pattern you associate with lasers. As such, they are having significant impact on the economics and performance of systems, especially in the area of high-speed networks.

PUMP LASERS

Pump lasers are characterized by their high output power at either 980 or 1480 nm. These lasers are used with the fiber amplifiers described in Chapter 12. They help amplify an optical signal without having first to convert it to an electrical signal. The pump laser

FIGURE 8–12
A VCSEL's output is circular in contrast to the elliptical output of an LED

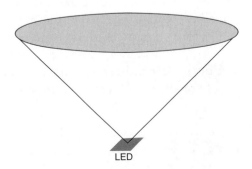

LED

VCSEL

supplies additional energy into a specially doped fiber and this energy is transferred to the signal, amplifying it.

Pump output powers can run over 100 mW. Other characteristics associated with sources are not important here. For example, the pump laser typically isn't modulated (turned on and off) since its job is to supply a constant optical energy and so operating speed is not important. Its output may be adjusted to maintain consistent output levels from the amplifier.

DIRECT vs INDIRECT MODULATION

In digital applications, the most common way to operate a laser or LED is by turning it on and off to form pulses. Actually, a laser is not turned on and off: its intensity is varied between a point near its threshold and a higher point. The reason is that the laser can operate very fast between two points above the threshold, but it is much slower at going from off to the lasing threshold. The point, though, is that the source is modulated by changing the input current between high and low levels. This type of modulation is called direct modulation. The output of the source is based on the input.

In very-high-speed applications—in the gigabit range—the light is modulated after it leaves the source. It becomes increasingly difficult to switch the laser between high and low levels at very high rates. It becomes easier to modulate the light emitted from the laser. The source emits a constant output and an external device is used to modulate the light. One way to do this is with a lithium niobate modulator. The source is connected by a length of fiber to the modulator. The electrical signal is also fed into the modulator. The modulator acts like a shutter, varying the intensity of the light in response to changes in the electrical signal.

Figure 8-13 shows direct and indirect modulation.

A newer type of externally modulated laser is the electroabsorption-modulated laser. The EM device has both the laser and modulator fabricated on a single chip as shown in Figure 8-14. The laser section and the modulator are isolated from one another. Designed primarily for long-distance applications, the laser is typically a distributed feedback device with a very narrow wavelength. EM lasers operating at 10 Gbit/s allow transmission distances of 1000 km without regeneration.

LED DRIVER CIRCUITS

LED optical output is approximately proportional to drive current. Other factors, such as temperature, also affect the optical output. This section shows three typical drive circuits for LEDs. It assumes you have some familiarity with electronic circuits.

FIGURE 8-13
Direct and indirect modulation

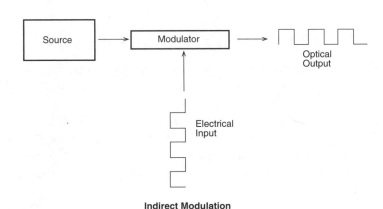

LEDs are usually driven with either a digital signal or an analog signal. When the drive signal is digital, there is no concern about LED linearity. The LED is either on or off. There are special problems that need to be addressed when designing an LED driver. The key concern is driving the LED so that maximum speed is achieved. Figure 8-15 shows three popular digital LED driver circuits. The first circuit, shown in Figure 8-15a, is a simple series driver circuit. The input voltage is applied to the base of transistor Q1 through resistor R1. The transistor is either off or on. When transistor Q1 is off, no current will flow through the LED, and no light will be emitted. When transistor Q1 is on, the cathode (bottom) of the LED will be pulled low. Transistor Q1 will pull its collector down to about 0.25 volts. The current is equal to the voltage across resistor R2, divided by the resistance of R2. The voltage across R2 is equal to the power supply voltage less the LED forward voltage drop and the saturation voltage of the drive transistor. The key advantage of the series driver shown in Figure 8-15a is its low average power supply current. If one defines the peak LED drive current as I_{LEDmax} and assumes that the LED duty cycle is 50%, then the average power supply current is only $I_{LEDmax}/2$. Further, the power dissipated is $(I_{LEDmax}/2)/V_{SUPPLY}$, where V_{SUPPLY} is the power supply voltage. The power dissipated by the individual components, the LED, transistor and resistor R1, is equal to the voltage drop across each component multiplied by $(I_{LEDmax}/2)$. The key disadvantage of

FIGURE 8-14
Electroabsorption-
modulated laser
(Courtesy of
Lucent
Technologies)

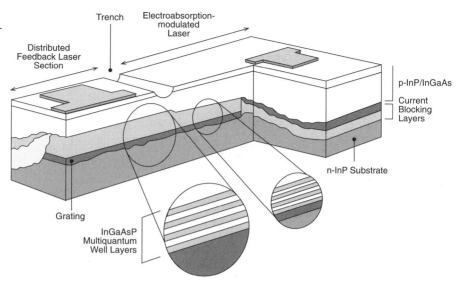

the circuit shown in Figure 8-15a is low speed. This type of driver circuit is rarely used at data rates above 30 to 50 Mbps. In general, there are two ways to design an LED drive circuit for low power dissipation. The first is to use a high-efficiency LED and reduce I_{LEDmax} to the lowest possible value. The second is to reduce the duty cycle of the LED to a low value. Usually larger gains can be made with the second method.

The second LED driver circuit, shown in Figure 8-15b, offers much higher speed capability. It uses transistor Q1 to quickly discharge the LED to turn it off. This circuit will drive the LED several times faster than the series drive circuit shown in Figure 8-15a. The key advantage of the shunt drive circuit is that it gives much better drive symmetry. LEDs are easy to turn on quickly, but are difficult to turn off because of the relatively long carrier lifetime. In the shunt driver circuit in Figure 8-15b, resistor R2 provides a positive current to turn on the LED. Typically, R2 would be in the 40 W range. This makes the turn-on current about 100 mA peak. Transistor Q1 provides the turnoff current. When saturated, transistor Q1 will have an impedance of a few ohms. This provides a much larger discharging current allowing the LED to turn off quickly. The key disadvantage of the shunt driver is the power dissipation. It is typically more than double that of the series driver. In fact, the circuit draws more current and power when the LED is off than when the LED is on! The exact power dissipation can be computed by first analyzing the off and on state currents and then combining the two values using information about the operating duty cycle.

The last driver circuit, shown in Figure 8-15c, is a variation on the shunt driver shown in Figure 8-15b. Two additional resistors and two capacitors have been added to the basic

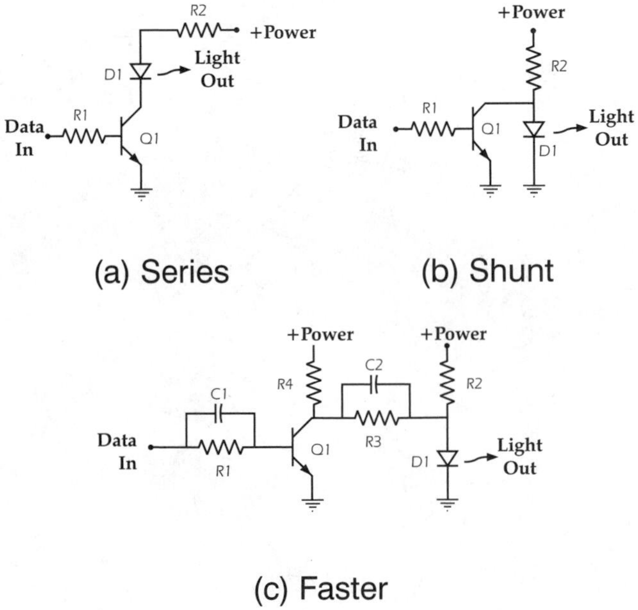

FIGURE 8-15
LED driver circuits (Courtesy of Force, Incorporated)

(a) Series   (b) Shunt

(c) Faster

circuit. The purpose of these additional components is to further improve the operating speed. Capacitor C1 serves to improve the turn-on and turn–off characteristics of transistor Q1 itself. One has to be careful that C1 is not made too large. If this occurs, the transistor base may be overdriven and damaged. The additional components, resistors R3 and R4 and capacitor C2, provide overdrive when the LED is turned on and underdrive when the transistor is turned off. The overdrive and underdrive accelerates the LED transitions. Typically, the RC time constant of R3 and C2 is made approximately equal to the rise or fall time of the LED itself when driven with a square wave. All of these tricks together can increase the operating speed of the LED and driver circuit to about 270 Mbps. There have been numerous laboratory tests and prototype circuits that have achieved rates to 500 to 1000 Mbps, but none of these have ever made it into mass production. Typically these levels of performance require a great deal of custom tweaking on each part to achieve the high data rates.

Lasers, of course, can operate much faster, but their drive circuits tend to be more complex because of the increased need to compensate for variations in temperature, drive current, and power supply voltage.

PACKAGING

A central issue in packaging the source is to couple as much optical power into the fiber as possible. The optical characteristics of the source—principally the output pattern—are not necessarily compatible with the fiber. To help couple more light into the fiber, some sort of lensing mechanism is often used to focus the light into the fiber. Figure 8-16 shows an example in which a small glass bead (micro lens) is used between the source and the fiber. Figure 8-17 shows a glass bead applied directly to the LED chip. Various other types of lenses are used, including holograms made from silicon. The hologram is a piece of silicon on which a pattern is etched. The hologram acts as a lens, again focusing the light into the fiber.

Another approach is the pigtail. Here, the fiber is permanently attached to the source package. A common approach is to cement the fiber into a V-groove and align the fiber and source to obtain optimum coupling. Used especially with lasers, pigtails bring the fiber up close to the actual laser chip so that the optical power is coupled into the fiber before it has a chance to spread out. The approach tends to be expensive since the alignment is made while the source is operated and the output from the fiber monitored. Newer approaches use precision parts that either eliminate or decrease the difficulty with which alignment is made.

FIGURE 8-16
Microlensed
LED and con-
nector (Courtesy
of Hewlett-
Packard)

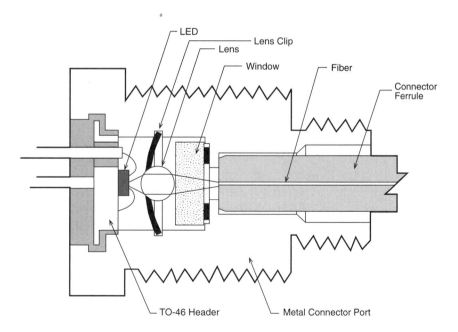

FIGURE 8-17
Microlens
applied directly
to chip (Illustra-
tion courtesy
of AMP
Incorporated)

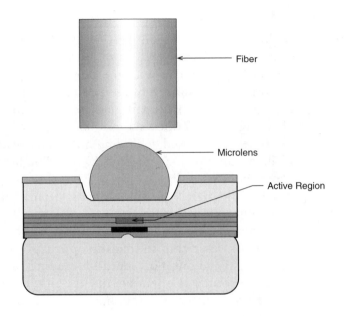

Fiber

Microlens

Active Region

The significance of source packaging can be seen in the offering of one manufacturer. The same LED chip is offered in three different packages. The specified output for each package is as follows:

- Unpigtailed
  300 µW
- Pigtailed with 100/140 fiber
  250 µW
- Pigtailed with 50/125 fiber
  50 µW

The difference in power in the two pigtailed versions results from the greater difficulty of coupling power into the small core and NA of the 50/125 fiber.

Sources are often packaged in a receptacle designed to mate with a specific type of connector. Figure 8-18 shows a variety of such receptacles. Such packaged solutions offer the user several benefits. First, the manufacturer can specify the amount of fiber coupled into a particular fiber size, which eliminates the need to calculate loss values for the source-to-fiber connection. Second, it simplifies design since the module is ready to mount on a PC board. Receptacles are available for popular connector types.

FIGURE 8-18
Source receptacle packages
(Courtesy of AMP Incorporated)

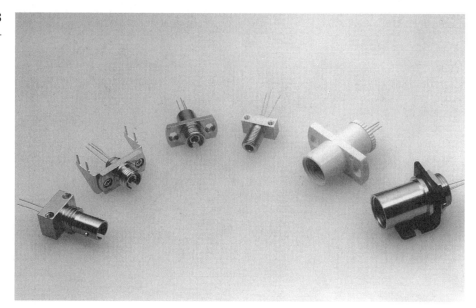

COOLING

Because of its small size and high output power, the laser can generate considerable heat. Heat has two effects on the laser. First, excessive heat can destroy the chip. Second, it can change the output by varying the wavelength output. The wavelength of a laser can drift 0.3 to 0.6 nm for every change of 1°C. Some applications carry signals on different wavelengths separated only by 0.8 nm. Wavelength variations can cause crosstalk, where one transmission becomes confused with another. A second consequence of changing wavelength is that the attenuation in a fiber also varies with wavelength. Wavelength stability is particularly important in applications carrying multiple wavelengths, such as the DWDM application discussed in Chapter 16.

If temperature control is important, the laser module typically has a cooling mechanism built in. One popular approach is a Peltier cooler, shown in Figure 8-19. Semiconductors are sandwiched between copper strips, which in turn are covered by a ceramic insulator. When an electrical polarity is applied to the device, heat will flow in the direction from positive to negative. Reversing the polarity reverses the direction of heat flow. The cooling is used to direct heat away from the laser and toward a heat sink that will absorb and dissipate the heat. Cooled lasers are often packaged in a relatively heavy metal package that acts as a heat sink. A thermoelectric cooler can maintain the laser at 15°C while the case temperature is 70°C.

FIGURE 8-19
Peltier cooler
(Courtesy of
Melcor
Corporation)

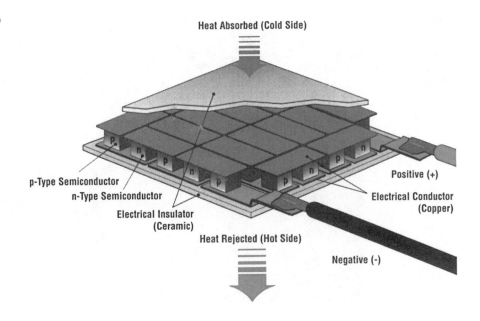

SAFETY

The optical energy emitted from LEDs and lasers is potentially dangerous to the eye. Lasers, in particular, are a concern because of their concentrated energy. Laser-based products are regulated by standards. In the United States, responsibility for these regulations is controlled by the Center for Devices and Radiological Health (CDRH) of the Food and Drug Administration. Outside of the United States, the principle regulation is IEC Publication 825. The two standards are generally similar in effect, although some differences exist. The regulations cover both the devices themselves and products using them.

Because a laser emits monochromatic, coherent, and highly collimated light, its energy is concentrated into a narrow beam. The energy density of this beam can harm biological tissues, particularly the eye. The chances of danger depend on the wavelength of the light, the amount of energy, and the time of exposure. The nature of the damage depends on the wavelength:

- Ultraviolet light (wavelengths under 400 nm): inflammation of the cornea and clouding of the lens.
- Visible and near-infrared light (400 to 1400 nm): thermal injury to retinal tissue. Fiber-optic systems typically operate in this range.
- Far-infrared light (greater than 1400 nm): damage to the cornea and lens. Many long-distance telephone applications operate at 1550 nm.

The CDRH and IEC regulations define four classes of devices as follows:

- **Class I:** These devices are considered inherently safe. The IEC requires a classification label, but the CDRH does not.
- **Class II:** Class 2 lasers have levels similar to a Class I device for an exposure of 0.25 second. Eye protection is normally provided by what is called a "normal aversion response." This means you involuntary blink.
- **Class III:** Both the CDRH and IEC define two subclasses: IIIa and IIIb. Class IIIa devices cannot injure a person's eye under normal conditions of bright light. They can, however, injure eyes when viewed through an optical aid such as a microscope or telescope. For Class IIIa, the CDRH concerns only visible light, while the IEC includes all wavelengths. Class IIIb devices can injure the eye if the light is viewed directly.
- **Class IV:** These devices are more powerful than even Class IIIb lasers. They can injure the eye even when viewed indirectly.

The regulations use equations to determine acceptable power levels at a given wavelength as well as procedures for making measurements or estimating power levels. A major difference between the CDRH and IEC regulations involves the method of testing. The CDRH requires the laser be tested under normal operating conditions. For the IEC, the device must be tested under fault conditions, such as if the laser is overdriven. Particularly with Class I lasers, this difference means that a laser that meets CDRH requirements may not meet IEC requirements. In addition, CDRH regulations consider only lasers; IEC requirements cover both lasers and LEDs. A third difference is that CDRH regulations are mandatory; IEC requirements are advisory. (A minor difference between the regulations: the CDRH uses roman numerals and the IEC uses arabic numerals.)

The output power allowed varies greatly with wavelength and the diameter of the beam. To give you an idea of the variety, Figure 8-20 gives CDRH levels for laser devices pigtailed to a single-mode optical fiber. Notice that Class I limits are much higher at 1550 nm than at 1310 nm. In short, the eye is more sensitive to lower wavelengths.

Most lasers in fiber optics are either Class I or Class IIIb devices. Class I devices require no special precautions. Class IIIb devices, besides cautionary labels and warnings in the documentation, require that circuits be designed to lessen the likelihood of accidental exposure. A safety interlock must be provided so that the laser will not operate if exposure is possible. One method is called open fiber control (OFC), which shuts down the laser if the circuit between the transmitter and receiver is open. A typical OFC system continuously monitors a link to ensure the link is operating intact by having the receiving circuit provide feedback to the transmitting circuit. If the receiving circuit does not receive data, the transmitting circuit stops operating the laser, under the assumption that a fault has occurred that might allow exposure to dangerous optical levels.

FIGURE 8-20
Acceptable power levels for CDRH classifications, with a single-mode pigtailed laser (Courtesy of Lucent Technologies)

WAVELENGTH	MODE FIELD DIAMETER	CLASS I MAXIMUM	CLASS IIIb RANGE
980 nm	6.6 μm	1 mW	1 mW–500 mW
1310 nm	8.8 μm	1.5 mW	1.5 mW–500 mW
1480 nm	8.8 μm	7.8 mW	7.8 mW–500 mW
1550 nm	8.8 μm	8.5 mW	8.5 mW–500 mW

Lasers must be labeled by classification. Because laser devices are very small, labeling each device is impractical. The packages they are shipped in, data sheets, and other related materials can be labeled, as can the equipment they are used in.

Regardless of regulations and laser classifications, it's a good practice to never look directly into a fiber or source that could be energized. Remember that most light used in fiber-optic systems is infrared energy that you cannot see—you will be exposed without noticing it. Your reaction can range from nothing (for very safe levels), to a protective blink or a damaged eye. Be especially careful when using an inspection microscope.

SOURCE EXAMPLES

Figure 8-21 provides typical specifications for LEDs and lasers. Figure 8-22 is a data sheet of a high-performance laser module.

FIGURE 8-21
Typical LED and laser characteristics

	LED MIN	LED TYP	LED MAX	LED MIN	LED TYP	LED MAX	LASER MIN	LASER TYP	LASER MAX
Coupled Power (μW)									
50-μm fiber	30	55	80	25	35	—	—	—	—
62.5-μm fiber	44	100	175	50	75	—	—	—	—
Single-mode fiber	—	—	—	0.5	1	—	1	1000	—
Wavelength (nm)	820	835	850	1290	1320	1350	1280	1310	1330
FWHM Spectral Width (nm)	—	—	75	—	—	170	—	—	5
Rise time (ns)	1.5	3.5	4	2	2.5	4	—	0.3	—

Data Sheet
January 1999

m i c r o e l e c t r o n i c s **group**

Lucent Technologies
Bell Labs Innovations

D370-Type Digital Uncooled
FastLight™ Laser Module

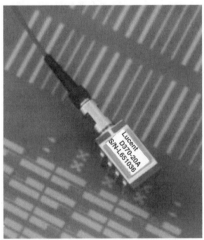

The low-profile D370-Type Laser Module is ideally suited for short- and long-reach SONET and other high-speed digital applications.

Features

- Eight-pin package suitable for SONET/SDH applications
- MQW F-P 1.3 µm laser with single-mode fiber pigtail
- Wide operating temperature range: −40 °C to +85 °C
- No TEC required

- High output power: typically 1.0 mW peak power coupled into single-mode fiber; 0.2 mW and 2.0 mW versions available
- Hermetically sealed active components
- Internal back-facet monitor
- Qualification program: Bellcore TA-983

Applications

- Long-reach SONET OC-3/OC-12 systems
- SDH STM-1/STM-4 systems
- Telecommunications
- Secure digital data systems

Benefits

- Easily board mounted
- Requires no lead bending
- No additional heat sinks required
- Pin compatible with industry-standard, 14-pin laser module

Description

The D370-Type Uncooled Laser Module consists of a laser diode coupled to a single-mode fiber pigtail. The device is available in a standard, 8-pin configuration (see Figure 1 and/or Table 1) and is ideal for long-haul (SONET) and other digital applications.

The module includes a multiquantum-well Fabry-Perot (MQW F-P) laser and an InGaAs PIN photodiode back-facet monitor in an epoxy-free, hermetically sealed package.

FIGURE 8-22

D370-Type Digital Uncooled *FastLight* Laser Module

**Data Sheet
January 1999**

Description (continued)

The device characteristics listed in this document are met at 1.0 mW output power. Higher- or lower-power operation is possible. Under conditions of a fixed photodiode current, the change in optical output is typically ±0.5 dB over an operating temperature range of –40 °C to +85 °C.

This device incorporates the new Laser 2000 manufacturing process from the Optoelectronic Products unit of Lucent Technologies Microelectronics Group. Laser 2000 is a low-cost platform that targets high-volume manufacturing and tight product distributions on all optical subassemblies. This platform incorporates an advanced optical design that is produced on Opto's highly automated production lines. The Laser 2000

platform is qualified for the central office and uncontrolled environments, and can be used for applications requiring high performance and low cost.

Table 1. Pin Descriptions

Pin Number	Connection
1	NC/Reserved
2	Case ground
3	NC/Reserved
4	Photodiode cathode
5	Photodiode anode
6	Laser diode cathode
7	Laser diode anode
8	NC/Reserved

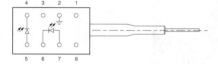

1-900

Figure 1. D370-Type Digital Uncooled Laser Module Schematic, Top View

Absolute Maximum Ratings

Stresses in excess of the absolute maximum ratings can cause permanent damage to the device. These are absolute stress ratings only. Functional operation of the device is not implied at these or any other conditions in excess of those given in the operations sections of the data sheet. Exposure to absolute maximum ratings for extended periods can adversely affect device reliability.

Parameter	Symbol	Min	Max	Unit
Maximum Peak Laser Drive Current or	I_{OP}	—	150	mA
Maximum Fiber Power*	P_{MAX}	—	10	mW
Peak Reverse Laser Voltage:				
Laser	V_{RL}	—	2	V
Monitor	V_{RD}	—	20	V
Monitor Forward Current	I_{FD}	—	2	mA
Operating Case Temperature Range	T_C	–40	85	°C
Storage Case Temperature Range	T_{stg}	–40	85	°C
Lead Soldering Temperature/Time	—	—	260/10	°C/s
Relative Humidity (noncondensing)	RH	—	85	%

* Rating varies with temperature.

2

FIGURE 8-22 continued

Data Sheet
January 1999   **D370-Type Digital Uncooled *FastLight* Laser Module**

Handling Precautions

Caution: **This device is susceptible to damage as a result of electrostatic discharge (ESD). Take proper pre-cautions during both handling and testing. Follow guidelines such as JEDEC Publication No. 108-A (Dec. 1988).**

Although protection circuitry is designed into the device, take proper precautions to avoid exposure to ESD.

Electro/Optical Characteristics

Table 2. Electro/Optical Characteristics (over operating temperature range unless otherwise noted)

Parameter	Symbol	Test Conditions	Min	Typ	Max	Unit
Operating Temperature Range	T	—	−40	—	85	°C
Optical Output Power*	P_F	CW, nominal	—	1	—	mW
Threshold Current	I_{TH}	T = 25 °C T = full range	5 2	9 —	15 45	mA mA
Modulation Current	I_{MOD}	CW, P_F = 1.0 mW, T = 25 °C CW, I_{MON} = constant, T = full range	10 8	15 —	20 35	mA mA
Slope Efficiency†	SE	CW, P_F = 1.0 mW, T = 25 °C	50	75	100	μW/mA
Center Wavelength	$λ_C$	P_F = 1.0 mW, CW	1270	—	1350	nm
RMS Spectral Width	Δλ	P_F = 1.0 mW, 155 Mbits/s	—	2	3	nm
Tracking Error	TE	I_{MON} = constant, CW	—	0.5	±1	dB
Spontaneous Emission	P_{TH}	I = I_{TH} x 0.9	—	—	50	μW
Rise/Fall Times	t_R, t_F	10%—90% pulse, T = 25 °C	—	0.25	0.5	ns
Forward Voltage	V_F	CW	—	1.1	1.6	V
Input Impedance	R	—	3	—	8	Ω
Monitor Current	I_{MON}	$V_R^‡$ = 5 V	150	—	750	μA
Monitor Dark Current	I_D	$V_R^‡$ = 5 V	—	10	200	nA
Wavelength Temperature Coefficient	—	—	—	0.4	0.5	nm/°C

* Higher and lower powers available. See Table 4 for more information.
† The slope efficiency is used to calculate the modulation current for a desired output power. This modulation current plus the threshold current comprise the total operating current for the device.
‡ V_R = reverse voltage.

FIGURE 8-22 continued

Characteristic Curve

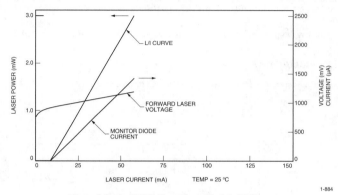

Figure 2. Typical D370-Type Laser Module L/I Curve

1-884

Lucent Technologies Inc.

FIGURE 8-22 continued

Data Sheet
January 1999   **D370-Type Digital Uncooled** *FastLight* **Laser Module**

Outline Diagram

Dimensions are in inches and (millimeters).

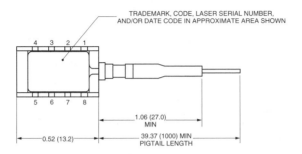

TRADEMARK, CODE, LASER SERIAL NUMBER,
AND/OR DATE CODE IN APPROXIMATE AREA SHOWN

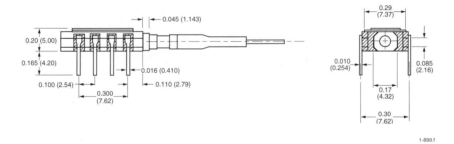

1-899.f

FIGURE 8-22 continued

Qualification Information

The D370-Type Laser Module has passed the following qualification tests and meets the intent of Bellcore TR-NWT-000468 for interoffice environments and TA-TSY-000983 for outside plant environments.

Table 3. D370-Type Laser Module Qualiÿcation Test Plan

Qualification Test	Conditions	Sample Size	Reference
Mechanical Shock	500 G	11	MIL-STD-883 Method 2002
Vibration	20 g, 20 Hz—2,000 Hz	11	MIL-STD-883 Method 2007
Solderability	—	11	MIL-STD-883 Method 2007
Thermal Shock	Delta T = 100 °C	11	MIL-STD-883 Method 2003
Fiber Pull	1 kg; 3 times	11	Bellcore 983
Accelerated (Biased) Aging	85 °C, 5,000 hrs.	25	Bellcore 983 Section 5.18
High-temperature Storage	85 °C, 2,000 hrs.	11	Bellcore 983
Temperature Cycling	500 cycles	11	Bellcore 983 Section 5.20
Cyclic Moisture Resistance	10 cycles	11	Bellcore 983 Section 5.23
Damp Heat	40 °C, 95% RH, 1344 hrs.	11	MIL-STD-202 Method 103
Internal Moisture	<5,000 ppm water vapor	11	MIL-STD-883 Method 1018
Flammability	—	—	TR357 Section 4.4.2.5
ESD Threshold	—	6	Bellcore 983 Section 5.22

Lucent Technologies Inc.

FIGURE 8–22 continued

Data Sheet
January 1999   D370-Type Digital Uncooled *FastLight* Laser Module

Laser Safety Information

Class IIIb Laser Product

This product complies with 21 CFR 1040.10 and 1040.11.
8.3 µm single-mode pigtail or connector
Wavelength = 1.3 µm
Maximum power = 10 mW
Because of size constraints, laser safety labeling is not affixed to the module but attached to the outside of the shipping carton.
Product is not shipped with power supply.

Caution: Use of controls, adjustments, and procedures other than those specified herein may result in hazardous laser radiation exposure.

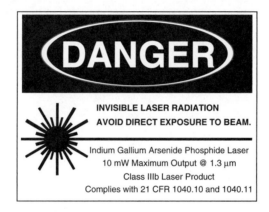

DANGER
INVISIBLE RADIATION IS EMITTED FROM THE END OF THE FIBER OR CONNECTOR. AVOID DIRECT EXPOSURE TO THE BEAM. DO NOT VIEW WITH OPTICAL INSTRUMENTS.

Lucent Technologies Inc.   7

FIGURE 8-22 continued

Ordering Information

Table 4. Ordering Information

Device Code	Comcode	Pfiber	Connector
D370-02A	107950859	0.2 mW	SC-PC
D370-10A	107950867	1.0 mW	SC-PC
D370-20A	107950875	2.0 mW	SC-PC
D370-10B	108096322	1.0 mW	SC-APC
D370-02F	107950883	0.2 mW	FC-PC
D370-10F	107950891	1.0 mW	FC-PC
D370-20F	107950909	2.0 mW	FC-PC
D370-02N	107950917	0.2 mW	none
D370-10N	107950925	1.0 mW	none
D370-20N	107950933	2.0 mW	none

For additional information, contact your Microelectronics Group Account Manager or the following:
INTERNET: **http://www.lucent.com/micro**, or for Optoelectronics information, **http://www.lucent.com/micro/opto**
E-MAIL: **docmaster@micro.lucent.com**
N. AMERICA: Microelectronics Group, Lucent Technologies Inc., 555 Union Boulevard, Room 30L-15P-BA, Allentown, PA 18103
                1-800-372-2447, FAX 610-712-4106 (In CANADA: **1-800-553-2448**, FAX 610-712-4106)
ASIA PACIFIC: Microelectronics Group, Lucent Technologies Singapore Pte. Ltd., 77 Science Park Drive, #03-18 Cintech III, Singapore 118256
                Tel. (65) 778 8833, FAX (65) 777 7495
CHINA: Microelectronics Group, Lucent Technologies (China) Co., Ltd., A-F2, 23/F, Zao Fong Universe Building, 1800 Zhong Shan Xi Road,
                Shanghai 200233 P. R. China   **Tel. (86) 21 6440 0468, ext. 316**, FAX (86) 21 6440 0652
JAPAN: Microelectronics Group, Lucent Technologies Japan Ltd., 7-18, Higashi-Gotanda 2-chome, Shinagawa-ku, Tokyo 141, Japan
                Tel. (81) 3 5421 1600, FAX (81) 3 5421 1700
EUROPE: Data Requests: MICROELECTRONICS GROUP DATALINE: **Tel. (44) 1189 324 299**, FAX (44) 1189 328 148
                Technical Inquiries: OPTOELECTRONICS MARKETING: **(44) 1344 865 900** (Ascot UK)

January 1999
DS99-033LWP (Replaces DS98-388LWP-01)

microelectronics group

Lucent Technologies
Bell Labs Innovations

FIGURE 8-22 continued

SUMMARY

- The source is the electro-optic transducer in which a forward current results in emission of light.
- An LED emits light spontaneously; a laser uses stimulated emission to achieve higher outputs above the threshold current.
- A laser has a narrower spectral width than an LED.
- A laser operates faster than an LED.
- Single-mode fibers require a laser.
- LEDs are less expensive and easier to operate than lasers.
- Source packaging is important to efficient coupling of light into a fiber.
- DFB lasers have ultra-narrow spectral widths to allow very-high-speed long distance communications.
- VCSELs offer circular outputs, low cost, and compatibility with multimode fibers

Review Questions

1. What is the purpose of the source?
2. Name the two main types of fiber-optic sources.
3. What is emitted when a free electron recombines with a hole in a semiconductor source?
4. From what phrase is the word *laser* derived?
5. List three packaging methods used to make coupling of light into a fiber more efficient.
6. If a source has a rise time of 7 ns, what is its approximate bandwidth?
7. List three characteristics of laser light that differentiate it from LED light.
8. What relationship between source NA and fiber NA results in NA mismatch loss?

chapter nine

DETECTORS

The *detector* performs the opposite function from the source: It converts optical energy to electrical energy. The detector is an optoelectronic transducer. A variety of detector types is available. The most common is the *photodiode,* which produces current in response to incident light. Two types of photodiodes used extensively in fiber optics are the pin photodiode and the avalanche photodiode. This chapter describes photodiode detectors and their characteristics most important to our study of fiber optics.

PHOTODIODE BASICS

Chapter 8 discussed the theory of energy bands in semiconductor material. In moving from the conduction band to the valence band, by recombination of electron-hole pairs, an electron gives up energy. In an LED, this energy is an emitted photon of light with a wavelength determined by the band gap separating the two bands. Emission occurs when current from the external circuit passes through the LED.

With a photodiode, the opposite phenomenon occurs: Light falling on the diode creates current in the external circuit. Absorbed photons excite electrons from the valence band to the conduction band, a process known as *intrinsic absorption*. The result is the creation of an electron-hole pair. These carriers, under the influence of the bias voltage applied to the diode, drift through the material and induce a current in the external circuit. For each electron-hole pair thus created, an electron is set flowing as current in the external circuit.

PN PHOTODIODE

The simplest photodiode is the pn photodiode shown in Figure 9-1. Although this type of detector is not widely used in fiber optics, it serves to illustrate the basic ideas of semiconductor photodetection. Other devices—the pin and avalanche photodiodes—are designed to overcome the limitations of the pn diode.

The pn photodiode is a simple pn device. When it is reverse biased (negative battery terminal connected to p-type material), very little current flows. The applied electric field creates a depletion region on either side of the pn junction. Carriers—free electrons and holes—leave the junction area. In other words, electrons migrate toward the negative terminal of the device (toward the positive terminal of the battery) and holes toward the positive terminal (negative terminal of the battery). Because the depletion region has no carriers, its resistance is very high, and most of the voltage drop occurs across the junction. As a result, electrical forces are high in this region and negligible elsewhere.

An incident photon absorbed by the diode gives a bound electron sufficient energy to move from the valence band to the conduction band, creating a free electron and a hole. If this creation of carriers occurs in the depletion region, the carriers quickly separate and drift rapidly toward their respective regions. This movement sets an electron flowing as current in the external circuit. When the carriers reach the edge of the depletion region, where electrical forces are small, their movement, and hence their external current, ceases.

FIGURE 9-1
Pn photodiode

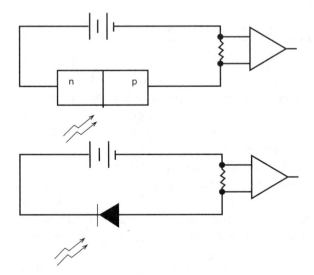

When electron–hole creation occurs outside of the depletion region, the carriers move slowly toward the depletion region. Many carriers recombine before reaching it. External currents are negligible. Those carriers remaining and reaching the depleted area are swiftly swept across the junction by the large electrical forces in the region to produce an external electrical current. This current, however, is delayed with respect to the absorption of the photon that created the carriers because of the initial slow movement of carriers toward the depletion region. Current, then, will continue to flow after the light is removed. This slow response, due to slow diffusion of carriers, is called *slow tail response.*

Two characteristics of the pn diode make it unsuitable for most fiber-optic applications. First, because the depletion area is a relatively small portion of the diode's total volume, many of the absorbed photons do not result in external current: The created holes and free electrons recombine before they cause external current. The received optical power must be fairly high to generate appreciable current. Second, the slow tail response from slow diffusion makes the diode too slow for medium- and high-speed applications. This slow response limits operations to the kilohertz range.

PIN PHOTODIODE

The structure of the *pin diode* is designed to overcome the deficiencies of its pn counterpart. The depletion region is made as large as possible by the pin structure shown in Figure 9-2. A lightly doped intrinsic layer separates the more heavily doped p-type and n-type materials. *Intrinsic* means that the material is not doped to produce n-type material with free electrons or p-type material with holes. Although the intrinsic layer is actually lightly doped positive, the doping is light enough to allow the layer to be considered intrinsic—that is, neither strongly n-type nor p-type. The name of the diode comes from this layering of materials: *p*ositive, *i*ntrinsic, *n*egative—pin.

Since the intrinsic layer has no free carriers, its resistance is high, and electrical forces are strong within it. The resulting depletion region is very large in comparison to the size of the diode. The pin diode works like the pn diode. The large intrinsic layer, however, means that most of the photons are absorbed within the depletion region for better efficiency. The result is improved efficiency in incident photons, creating external current and faster speed. Carriers created within the depletion region are immediately swept by the electric field toward their p or n terminals.

A tradeoff exists in arriving at the best pin photodiode structure. Efficiency of photon-to-carrier conversion requires a thick intrinsic layer to increase the probability of incident photons creating electron-hole pairs in the depletion region. Speed, however, requires a thinner layer to reduce the transit time of the carriers swept across the region. Diode design involves balancing these opposing requirements to achieve the best balance between efficiency and speed.

FIGURE 9-2
Pin photodiode
(Illustration
courtesy of AMP
Incorporated)

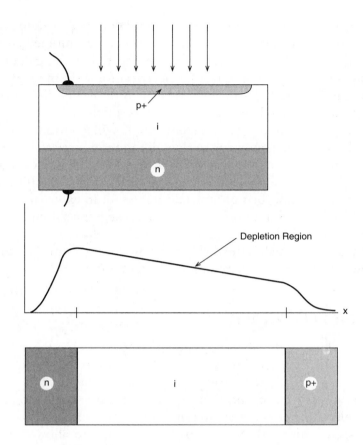

AVALANCHE PHOTODIODE (APD)

For a pin photodiode, each absorbed photon ideally creates one electron-hole pair, which, in turn, sets one electron flowing in the external circuit. In this sense, we can loosely compare it to an LED. There is basically a one-to-one relationship between photons and carriers and current. Extending this comparison allows us to say that an avalanche photodiode resembles a laser, where the relationship is not one-to-one. In a laser, a few primary carriers result in many emitted photons. In an avalanche photodiode (APD), a few incident photons result in many carriers and appreciable external current.

The structure of the APD, shown in Figure 9-3, creates a very strong electrical field in a portion of the depletion region. *Primary carriers*—the free electrons and holes created by absorbed photons—within this field are accelerated by the field, thereby gaining several electron volts of kinetic energy. A collision of these fast carriers with neutral atoms causes the accelerated carrier to use some of its energy to raise a bound electron from the

FIGURE 9-3
Avalanche pho-
todiode
(Illustration
courtesy of AMP
Incorporated)

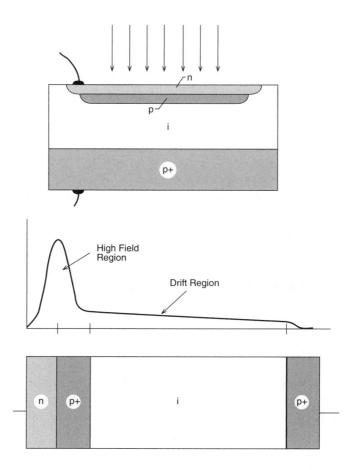

valence band to the conduction band. A free electron and hole appear. Carriers created in this way, through collision with a primary carrier, are called *secondary carriers.*

This process of creating secondary carriers is known as *collision ionization.* A primary carrier can create several new secondary carriers, and secondary carriers themselves can accelerate and create new carriers. The whole process is called *photomultiplication,* which is a form of gain.

The number of electrons set flowing in the external circuit by each absorbed photon depends on the APD's multiplication factor. Typical multiplication ranges in the tens and hundreds. A multiplication factor of 70 means that, on the average, 70 external electrons flow for each photon. The phrase "on the average" is important. The multiplication factor is an average, a statistical mean. Each primary carrier created by a photon may create more or less secondary carriers and therefore external current.

For an APD with a multiplication factor of 70, for example, any given primary carrier may actually create 67 secondary carriers or 76 secondary carriers. This variation is one source of noise that limits the sensitivity of a receiver using an APD. We will discuss noise in fuller detail shortly.

The multiplication factor varies with the bias voltage. Because the accelerating forces must be strong enough to impart energies to the carriers, high bias voltages (several hundred volts in many cases) are required to create the high-field region. At lower voltages, the APD operates like a pin diode and exhibits no internal gain.

The avalanche breakdown voltage of an APD is the voltage at which collision ionization begins. An APD biased above the breakdown point will produce current in the absence of optical power. The voltage itself is sufficient to create carriers and cause collision ionization.

The APD is often biased just below the breakdown point, so any optical power will create a fast response and strong output. The tradeoffs are that dark current (the current resulting from generation of electron-hole pairs even in the absence of absorbed photons) increases with bias voltage, and a high-voltage power supply is needed.

NOISE

The purpose of the detector is to create an electrical current in response to incident photons. It must accept highly attenuated optical energy and produce a current. This current is usually feeble because of the low levels of optical power involved, often only tens or hundreds of nanowatts. Subsequent stages of the receiver amplify and possibly reshape the signal from the detector.

Noise is an ever-present phenomenon that seriously limits the detector's operation. *Noise* is any electrical or optical energy apart from the signal itself. The signal is wanted energy; noise is anything else—that is, unwanted energy. Although noise can and does occur in every part of a communication system, it is of greatest concern in the receiver input. The reason is that the receiver works with very weak signals that have been attenuated during transmission. Although very small compared to the signal levels in most circuits, the noise level is significant in relation to the weak detected signals. The same noise level in a transmitter is usually insignificant because signal levels are very strong in comparison. Indeed, the very limit of the diode's sensitivity is the noise. An optical signal that is too weak cannot be distinguished from the noise. To detect such a signal, we must either reduce the noise level or increase the power level of the signal.

The amplification stages of the receiver amplify both the signal and the noise. Although it is possible to use electronic circuits to filter out some types of noise, it is better to have the signal much stronger than the noise by either having a strong signal level or a low noise level. Having both is even better.

Several types of noise are associated with the photodetector itself and with the receiver. We have already mentioned multiplication noise in an APD, which arises because multiplication varies around a statistical mean. Two other types of noise most important to our understanding of photodiodes and fiber optics are shot noise and thermal noise.

SHOT NOISE

Shot noise arises from the discrete nature of electrons. Current is not a continuous, homogeneous flow. It is the flow of individual, discrete electrons. Keep in mind that a photodiode works because an absorbed photon creates an electron-hole pair that sets an external electron flowing as current. It is a three-step sequence: photon, electron-hole carriers, electron. The arrival and absorption of each photon and the creation of carriers are part of a random process. It is not a perfect homogeneous stream but a series of discrete occurrences. Therefore, the actual current fluctuates, as more or less electron-hole pairs are created in any given moment.

Shot noise exists even when no light falls on the detector. Even without light, a small trickle of current is generated thermally, increasing about 10% for every increase of 1°C. A typical dark current is 25 nA at 25°.

Shot noise is equal to

$$i_{sn}^2 = 2qiB$$

where q is the charge of an electron (1.6×10^{-19} coulomb), i is average current (including dark current and signal current), and B is the receiver bandwidth. The equation shows that shot noise increases with current and with bandwidth. Shot noise is at its minimum when only dark current exists (when i = dark current), and it increases with the current resulting from optical input. A detector with a dark current of 2 nA and operating at a bandwidth of 10 MHz has a shot noise of 80 pA:

$$
\begin{aligned}
i_{sn}^2 &= 2qiB \\
&= (2)(1.6 \times 10^{-19})(2 \times 10^{-9})(10 \times 10^{6}) \\
&= 6.4 \times 10^{-21} \\
i_{sn} &= 8 \times 10^{-11} \\
&= 80 \text{ pA}
\end{aligned}
$$

THERMAL NOISE

Thermal noise, also called *Johnson noise* or *Nyquist noise,* arises from fluctuations in the load resistance of the detector. The electrons in the resistor are not stationary: Their thermal energy allows them to constantly and randomly move around. At any given instant, the net movement can be toward one electrode or the other, so that randomly varying current exists. This random current will add to and distort the signal current from the photodiode.

Thermal noise is equal to

$$i_{tn}{}^2 = \frac{4kTB}{R_L}$$

where k is Boltzmann's constant (1.38×10^{-23} J/K), T is absolute temperature (kelvin scale), B is the receiver's bandwidth, and R_L is the load resistance.

Consider a 510-Ω load resistance operating at an absolute temperature of 298 K. Assume a bandwidth of 10 MHz. The thermal noise is

$$
\begin{aligned}
i_{tn}{}^2 &= \frac{4kTB}{R_L} \\
&= \frac{(4)(1.38 \times 10^{-23})(298)(10 \times 10^6)}{510} \\
&= 3.23 \times 10^{-16} \\
i_{tn} &= 1.79 \times 10^{-8} \\
&= 18 \text{ nA}
\end{aligned}
$$

Thermal and shot noise exist in the receiver independently of the arriving optical power. They result from the very structure of matter. They can be minimized by careful design of devices and circuits, but they cannot be eliminated. Any signal—optical, electrical, or human—must exist in the presence of noise. In the receiver stages following the detector, the signal and noise will be amplified. Therefore, the signal must be appreciably larger than the noise. If the signal power is equal to the noise power, the signal will not even be adequately detected. As a general rule, the optical signal should be twice the noise current to be adequately detected.

SIGNAL-TO-NOISE RATIO

Signal-to-noise ratio (SNR) is a common way of expressing the quality of signals in a system. SNR is simply the ratio of the average signal power to the average noise power from all noise sources.

$$\text{SNR} = \frac{S}{N}$$

In decibels, SNR equals

$$\text{SNR} = 10 \log_{10} \left(\frac{S}{N} \right)$$

If the signal current is 50 μW and the noise power is 50 nW, the ratio is 1000, or 30 dB.

A large SNR means that the signal is much larger than the noise. The signal power depends on the power of the arriving optical power. Different applications require different SNRs—that is, different levels of "fidelity" or freedom from distortion. The SNR required for a telephone voice channel is less than that required for a television signal, since a fair amount of noise on a telephone line will go unnoticed. We will also accept greater distortion in voices than we will accept in television picture quality. Further, a broadcast-quality television signal, which is the signal produced by the television networks, has a higher SNR than the television signal received in our homes. Why? The broadcast signal itself encounters noise during its transmission. It must, therefore, have a higher SNR so that the signal, after picking up noise during its transmission and reception, still has an SNR high enough to produce a clear, sharp picture.

BIT-ERROR RATE

For digital systems, bit-error rate (BER) usually replaces SNR as a measure of system quality. BER is the ratio of incorrectly transmitted bits to correctly transmitted bits. A ratio of 10^{-9} means that one wrong bit is received for every 1 billion bits transmitted. As with SNR, the required BER varies with the application. Digitally encoded telephone voices have a lower BER requirement than digital computer data, say 10^{-6} versus 10^{-9}. A few faulty bits will not cause noticeable distortion of a voice. A few faulty bits in computer data can cause significant changes in financial data or student grades. They can mean the difference between a program running successfully or crashing.

BER and SNR are related. A better SNR brings a better BER. BER, though, also depends on data-encoding formats and receiver designs. Techniques exist to detect and correct bit errors. We cannot easily calculate the BER from the SNR, because the relationship depends on several factors, including circuit design and bit-error correction schemes. In a given system, for example, an SNR of 22 dB may be required to maintain a BER of 10^{-9}, whereas an SNR of 17 dB brings a BER of 10^{-6}. In another design, however, the 10^{-9} BER might be achieved with an SNR of 18 dB.

DETECTOR CHARACTERISTICS

The detector characteristics of interest here are those that relate most directly to use in a fiber-optic system. These include the device's response to incident optical power and its speed. Since pin diodes are the most commonly used, we concentrate on those.

Responsivity

Responsivity (R) is the ratio of the diode's output current to input optical power and is given in amperes/watt (A/W). Optical power produces current. A pin photodiode typically has a responsivity of around 0.4 to 0.6 A/W. A responsivity of 0.6 A/W means that incident light having 50 μW of power results in 30 μA of current:

$$I_d = 50 \ \mu W \times 0.6 \ A/W = 30 \ \mu A$$

where I_d is the diode current.

For an APD, a typical responsivity is 75 A/W. The same 50 μW of optical power now produces 3.75 mA of current.

$$I_d = 50 \ \mu W \times 75 \ A/W = 3750 \ \mu A = 3.75 \ mA$$

Responsivity varies with wavelength, so it is specified either at the wavelength of maximum responsivity or at a wavelength of interest, such as 850 nm or 1300 nm. Silicon is the most common material used for detectors in the 800- to 900-nm range. Its peak responsivity is 0.7 A/W at 900 nm. At the 850-nm wavelength, responsivity is still near the peak. In a fiber-optic system using plastic fibers, operation is usually at 650 nm in the visible spectrum. Here, the photodiode's responsivity drops to around 0.3 to 0.4 A/W.

Silicon photodiodes are not suitable for the longer wavelengths of 1300 nm and 1550 nm. Materials for long wavelengths are principally germanium (Ge) and indium gallium arsenide (InGaAs). An InGaAs pin photodiode has a fairly broad and flat response curve. It does not peak nearly as sharply as does silicon's. Its responsivity of greater than 0.5 A/W from 900 to 1650 nm allows it to be used at both 1300 nm and 1550 nm. Figure 9-4 depicts typical responsivity curves for photodiodes.

FIGURE 9-4
Responsivity

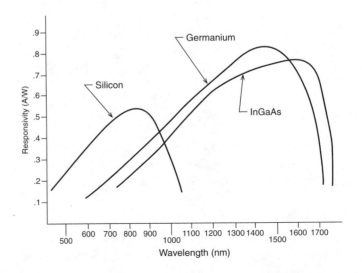

Quantum Efficiency (η)

Quantum efficiency is the ratio of primary electron-hole pairs (created by incident photons) to the photons incident on the diode material. It is expressed either as a dimensionless number or as a percentage. A quantum efficiency of 1, or 100%, means that every absorbed photon creates an electron-hole pair. A typical quantum efficiency of 70% means that only 7 out of every 10 photons create carriers (electron current). Quantum efficiency applies to primary electrons created by photon absorption, not to secondary carriers created by collision ionization.

Quantum efficiency deals with the fundamental efficiency of the diode for converting photons into free electrons. Responsivity can be calculated from quantum efficiency:

$$R = \frac{\eta e \lambda}{hc}$$

where e is the charge of an electron, h is Planck's constant, and c is the velocity of light. Since e, c, and h are constants, responsivity is simply a function of quantum efficiency and wavelength.

Dark Current

As mentioned earlier, *dark current* is the thermally generated current in a diode. It is the lowest level of thermal noise. Dark current increases about 10% for every increase of 1°C. It is much lower in Si photodiodes used at shorter wavelengths than in Ge or InGaAs photodiodes used at longer wavelengths.

Minimum Detectable Power

The minimum power detectable by the detector determines the lowest level of incident optical power that the detector can handle. In simplest terms, it is related to the dark current in the diode, since the dark current will set the lower limit. Other noise sources are factors, including those associated with the diode and those associated with the first stage of the receiver.

The *noise floor* of a pin diode, which tell us the minimum detectable power, is the ratio of noise current to responsivity:

$$\text{Noise floor} = \frac{\text{noise}}{\text{responsivity}}$$

For initial evaluation of a diode, we can use the dark current to estimate the noise floor. Consider a pin diode with $R = 0.5\ \mu A/\mu W$ and a dark current of 2 nA. The minimum detectable power is

$$\text{Noise floor} = \frac{2 \times 10^{-12}\ A}{0.5\ \mu A/\mu W}$$

$$= 4 \times 10^{-12}\ W$$

$$= 4\ nW$$

More precise estimates must include other noise sources, such as thermal and shot noise. As discussed, the noise depends on current, temperature, load resistance, and bandwidth.

Response Time

Response time is the time required for the photodiode to respond to optical inputs and produce external current. As with a source, response time is usually specified as a rise time and a fall time, measured between the 10% and 90% points of amplitude. Rise times range from 0.5 ns to tens of nanoseconds. Rise time is limited to the transit speed of the carriers as they sweep across the depletion region. It is influenced by the bias voltage: higher voltages bring faster rise times. A pin diode, for example, might have a rise time of 5 ns at 15 V and 1 ns at 90 V.

The response time of the diode relates to its usable bandwidth. Bandwidth can be approximated from rise time:

$$BW = \frac{0.35}{t_r}$$

The bandwidth, or operating range, of a photodiode can be limited by either its rise time or its RC time constant, whichever results in the slower speed or bandwidth. The bandwidth of a circuit limited by the RC time constant is

$$BW = \frac{1}{2\pi R_L C_d}$$

where R_L is the load resistance and C_d is the diode capacitance. The rise time of the circuit is

$$t_r = 2.19\ R_L C_d$$

Figure 9-5 shows the equivalent circuit model of a pin diode. It consists of a current source in parallel with a resistance and a capacitance. It appears as a low-pass filter, a resistor-capacitor network that passes low frequencies and attenuates high frequencies. The *cutoff* frequency, which is the frequency that is attenuated 3 dB or 50%, marks the bandwidth. Frequencies higher than cutoff are eliminated.

FIGURE 9-5
Electrical model
of a pin diode
(Illustration
courtesy of AMP
Incorporated)

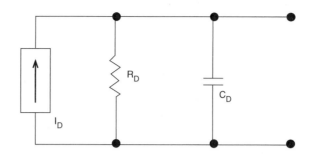

Diodes for high-speed operation must have capacitances of only a few picofarads or less. The capacitance in a pin diode is mainly the junction capacitance formed at the junctions of p, i, and n layers, as well as any capacitance contributed by the packaging structure and mounting.

Consider a photodiode with a rise time of 1 ns and a capacitance of 2 pF. Its approximate bandwidth, based on rise time, is

$$BW = \frac{0.35}{1 \text{ ns}}$$

$$= 0.35 \text{ GHz} = 350 \text{ MHz}$$

To ensure that the RC time constant does not lower the bandwidth, we must determine the highest resistor value usable:

$$BW = \frac{1}{2\pi R_L C_d}$$

$$350 \times 10^6 \text{ Hz} = \frac{1}{(2)(3.1415927)(2 \times 10^{-12})(R_L)}$$

$$R_L = 227 \text{ } \Omega$$

Thus, a standard 220-Ω resistor works, although in practice a resistor with one quarter of this value is often chosen.

Bias Voltage

Photodiodes require bias voltages ranging from as low as 5 V for some pin diodes to several hundred volts for APDs. Bias voltage significantly affects operation, since dark current, responsivity, and response time all increase with bias voltage. APDs are usually biased near their avalanche breakdown point to ensure fast response.

FIGURE 9-6
Effects of bias voltage and temperature on responsivity of an APD (Illustration courtesy of AMP Incorporated)

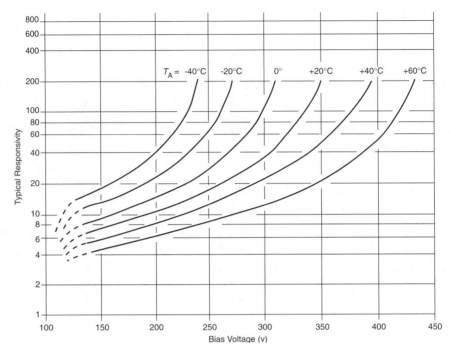

As shown in Figure 9-6, the APD is also sensitive to variations in temperature. The bias voltage required to maintain a given responsivity varies significantly with temperature. The output from an APD becomes erratic unless extensive compensating circuitry is employed in the receiver. The high voltage requirement and temperature sensitivity mean that APDs are chosen only when their responsivity and gain justify their complex application requirements.

INTEGRATED DETECTOR/PREAMPLIFIER

The *integrated detector/preamplifier* (IDP) is an alternative to pin photodiodes. Noise that limits receiver operation can occur between the diode itself and the first receiver stage. The electrical leads of the diode may be sensitive to surrounding EMI and act as an antenna, picking up noise that will be coupled along with the noise into the amplifier.

To reduce these noise sources, the IDP has a *transimpedance amplifier* incorporated on the same semiconductor chip as the photodetector. The transimpedance amplifier performs both amplification and current-to-voltage conversion. The IDP, then, is an integrated circuit having both a photodiode and a transimpedance amplifier. (The amplifier

FIGURE 9-7
Integrated detector/preamplifier (Illustration courtesy of AMP Incorporated)

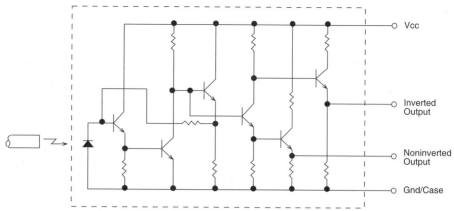

can also be placed in the detector package as a separate chip.) Figure 9-7 shows a schematic of an integrated IDP.

Characteristic specifications for an IDP are similar to those for regular photodetectors. The main difference is that the IDP's output is a voltage, so responsivity is specified in volts/watt (V/W). Since the IDP exhibits gain, typical responsivities are on the order of 40 V/W. For a 50-μW optical input, the output of the IDP would be 2 mV.

An alternative approach is a nonintegrated detector/preamp. Here a detector and separate preamplifier are included in the same package. The advantage is the same as for the integrated device: to have the two units close together to avoid noise problems.

PACKAGING

The packaging of the detector is similar to that of the source: TO series cans, pigtailed devices, and microlensed devices predominate. See the discussion of source packaging in Chapter 8 for further details.

As with sources, two main sources of loss in coupling light from a fiber into the detector results from mismatches in diameter and NA: When $\mathrm{dia_{det}} < \mathrm{dia_{fiber}}$,

$$\mathrm{loss_{dia}} = 10 \, \log_{10} \left(\frac{\mathrm{dia_{det}}}{\mathrm{dia_{fiber}}} \right)^2$$

When $\mathrm{NA_{det}} < \mathrm{NA_{fiber}}$,

$$\mathrm{loss_{NA}} = 10 \, \log_{10} \left(\frac{\mathrm{NA_{det}}}{\mathrm{NA_{fiber}}} \right)^2$$

FIGURE 9-8
Typical detector characteristics

	PIN	PIN/PREAMP	APD
Responsivity	.80 μA/μW	2 mV/μW	70 μA/μW
Spectral Response (nm)	1150–1600	1150–1600	1150–1600
Dark Current (nA)	2		5
Capacitance (pF)	1.5		4
Rise time (ns)	0.5 max		0.5

Since detectors can be easily manufactured with large active diameters and wide angles of view, such mismatches are less common than with sources. Other losses occur from Fresnel reflections and mechanical misalignment between the connector and diode package.

Detectors are packaged in the same receptacles as sources (see Figure 8-18).

DETECTOR SPECIFICATIONS

Figure 9-8 provides typical specifications for typical pin photodiodes and APDs.

SUMMARY

- The detector is the optoelectronic transducer that converts optical power into current.
- The two most common detectors used in fiber-optic communications are the pin photodiode and avalanche photodiode.
- Responsivity expresses the ratio of output current to optical power.
- An APD provides internal gain, so its responsivity is much higher than that of a pin diode. The APD is also more difficult to apply.
- Noise sets the lower limit of detectable optical power.
- Two types of noise associated with a detector are shot noise and thermal noise.
- SNR and BER are two methods of expressing the "quality" of a signal in a system.
- The response speed of a detector can be limited by its rise time or by its RC time constant.
- An IDP is a detector package containing both a pin photodiode and a transimpedance amplifier.

Review Questions

1. What is the purpose of the detector?
2. Name the two types of fiber-optic sources.
3. Name the three layers of a pin photodiode. What is the purpose of the middle layer?
4. Name the type of noise that results from current flowing as discrete electrons.
5. Name the type of noise that results from thermal energy in the load resistor.
6. Calculate the current that results from a received optical power of 750 nW and a responsivity of 0.7 A/W.
7. Name two factors that limit the response speed of a detector.
8. Calculate the rise time of a detector circuit having a capacitance of 3 pF and a load resistance of 160 Ω.
9. What distinguishes an IDP from a pin photodiode?
10. Give two reasons why a pin photodiode is used more often than an APD.

chapter ten

TRANSMITTERS AND RECEIVERS

In Chapter 1, we saw that a fiber-optic link contains a transmitter, optical cable, and receiver. We have looked at fiber-optic cables, sources, and detectors, the electro-optic transducers that provide the bridge between the optical and electronic parts of a fiber-optic system.

BASIC TRANSMITTER CONCEPTS

Figure 10-1 shows a basic block diagram for an LED-based transmitter. The transmitter typically connects to an application-specific chip that takes data and prepares it for the transmitter. Functions include accepting parallel data and converting it to a serial stream, encoding it for transmission, and sending it to the transmitter. The transmitter's input buffer accepts this data and provides an output current to drive the LED. The bias generator helps to maintain proper source output by compensating for variations in temperature and power supply voltage. Without such compensation, the transmitter's output could vary significantly with changes in temperature or supply voltage. For example, over a temperature range from 0 to 80°C, the output of a transmitter without compensation can vary by as much as 3 dB. With temperature compensation, variations can be held to under 0.5 dB.

A laser transmitter is more complex, as shown in Figure 10-2. Remember that a laser is not turned on and off: It is simply modulated between high and low levels above the threshold current. What's more, smaller variations in current produce large changes in the laser's output. Finally, the laser's threshold current and its output versus current can vary significantly with changes in temperature. To maintain a constant output over a range of temperatures and current variations, the laser transmitter has additional circuitry. The output from the rear of the laser drives a monitor photodiode. The photodiode feeds though a

FIGURE 10-1
LED-based
transmitter

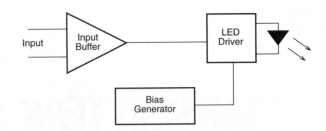

signal conditioner to the bias generator. The bias generator adjusts the drive current to the laser so that the output of the photodiode remains constant. If the photodiode's output is constant, so is the output of the transmitter. The bias generator is also fed by a reference generator that helps compensate for variations in temperature. A duty-cycle compensation circuit can also be added.

LOGIC FAMILIES

Most electronic systems operate on standard, well-defined signal levels. Television video signals use a 1-V peak-to-peak level. Digital systems use different standards, depending on the type of logic circuits used in the system. These logic circuits define the levels for the highs and lows that represent the 1s and 0s of digital data.

Currently, the most common digital logic is transistor-transistor logic (TTL). TTL uses 0.5 V for a low signal and 5 V for a high signal. Although TTL is a fast logic, allowing it to be used in many applications, an even faster logic is emitter-coupled logic (ECL). ECL circuits use −1.75 V for low and −0.9 V for high. Figure 10-3 shows the logic levels for TTL and ECL circuits.

FIGURE 10-2
Laser-based
transmitter

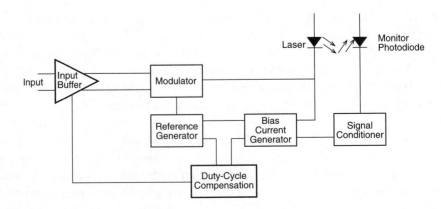

FIGURE 10-3
TTL and ECL
signal levels

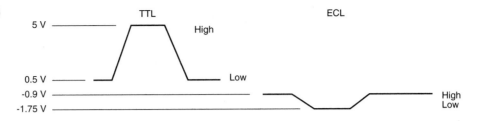

Normally, TTL and ECL logics cannot be intermixed; however, special chips allow signals to be converted from one logic level to another. Both TTL and ECL are used in fiber-optic transmitters and receivers. TTL is the more common; ECL is usually found only in high-speed systems—those above 50 or 100 Mbps.

An important and popular variation of ECL is positive or pseudo-ECL, which is called PECL. PECL uses the same +5-volt power supply as TTL and other circuits. Since most applications do not use ECL outside of the fiber-optic transmitter and receiver, the ability to use a common supply voltage simplifies design.

Another very popular logic family is complementary metal-oxide semiconductor (CMOS). It is rapidly becoming a popular replacement for TTL because of its very low power consumption. Many CMOS circuits use the same voltage levels as TTL for their signals.

The drive circuit of the transmitter must accept these logic levels. It then provides the output current to drive the source. For example, it may convert the 0.5 V and 5 V of TTL into 0 mA and 50 mA to turn the source on and off.

An additional function of the transmitter is to produce the proper modulation code.

MODULATION CODES

A string of digital highs and lows is often not suitable for transmission over any appreciable distance. Whereas digital circuits use simple high and low pulses to represent 1s and 0s of binary data, a more complex format is often used to transmit digital signals between electronic systems. A modulation code is a method of encoding digital data for transmission.

Earlier, we saw that the pulse represents a binary 1, and the absence of a pulse represents a binary 0. In a TTL system, a 5-V high represents a 1; a 0.5-V low represents a 0. Thus, there is a one-to-one correspondence between a high or low voltage and a binary 1 or 0. With the modulation codes described here, this correspondence does not always exist.

Figure 10-4 shows several popular modulation codes. Each bit of data must occur with its bit period, which is defined by the clock. The *clock* is a steady string of pulses that provides basic system timing. As shown in the table in Figure 10-4, some codes are self-

FIGURE 10-4
Modulation codes

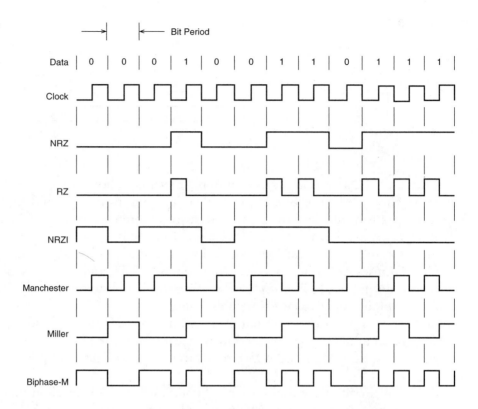

Format	Symbols Per Bit	Self-Clocking	Duty Factor Range (%)
NRZ	1	No	0 - 100
RZ	2	No	0 - 50
NRZI	1	No	0 - 100
Manchester (Biphase-L)	2	Yes	50
Miller	1	Yes	33 - 67
Biphase-M (Bifrequency)	2	Yes	50

clocking, and others are not. A self-clocking code means that the clock information is contained within the code. In a non-self-clocking system, this timing information is not present.

Clock information is important to the receiver. One purpose of a receiver is to rebuild signals to their original state as they were presented to the transmitter. To do so, the receiver must know the timing information. There are three alternatives:

1. The transmitted information can also contain clock information; in other words, the modulation code is self-clocking.

2. The clock or timing information must be transmitted on another line. This, of course, adds to system complexity by increasing the number of lines from transmitter to receiver. In a long-distance system, such costs may be considerable.
3. The receiver may provide its own timing and not rely on clock signals from the transmitter.

NRZ Code

The *NRZ (nonreturn-to-zero) code* is similar to "normal" digital data. The signal is high for a 1 and low for a 0. For a string of 1s, the signal remains high. For a string of 0s, it remains low. Thus, the level changes only when the data level changes.

RZ Code

The *RZ (return-to-zero) code* remains low for 0s. For a binary 1, the level goes high for one half of a bit period and then returns low for the remainder. For each 1 of data, the level goes high and returns low within each bit period. For a string of three 1s, for example, the level goes high for each 1 and returns to low.

NRZI Code

In an *NRZI (nonreturn-to-zero, inverted) code,* a 0 is represented by a change in level, and a 1 is represented by no change in level. Thus, the level will go from high to low or from low to high for each 0. It will remain at its present level for each 1. An important thing to notice here is that there is no firm relationship between 1s and 0s of data and the highs and lows of the code. A binary 1 can be represented by either a high or a low, as can a binary 0.

Manchester Code

A *Manchester code* uses a level transition in the middle of each bit period. For a binary 1, the first half of the period is high, and the second half is low. For a binary 0, the first half is low, and the second half is high.

Miller Code

In the *Miller code,* each 1 is encoded by a level transition in the middle of the bit period. A 0 is represented either by no change in level following a 1 or by a change at the beginning of the bit period following a 0.

Biphase-M Code

In the *bi-phase-M code,* each bit period beings with a change of level. For a 1, an additional transition occurs in midperiod. For a 0, no additional change occurs. Thus, a 1 is at both high and low during the bit period. A 0 is either high or low, but not both, during the entire bit period.

nB/nB ENCODING

Most low-speed networks use Manchester or differential Manchester encoding. At higher speeds, Manchester encoding falls from favor because it requires a clock rate twice that of the data rate. A 100-Mbps network requires a 200-MHz clock. NRZI data transmissions have no transitions when all zeros are present, which eliminates self-clocking.

High-speed networks use a group encoding scheme in which data bits are encoded into a data word of longer bit length. For example, the 4B/5B method encodes four data bits into a 5-bit code word. The receiver decodes the 5-bit word into four bits. This scheme guarantees that the data never have more than three consecutive 0s. Such encoding requires much less bandwidth than Manchester encoding. The overhead is only 20%. Thus high-speed networks using 4B/5B, such as FDDI or Fast Ethernet, transmit at a 125-Mbps rate, but the data rate is 100 Mbps. The 25-Mbps difference is due to the fifth bit in the 4B/5B encoding. Even with 4B/5B, the data is still transmitted with one of the encoding schemes in Figure 10-4. Other encoding schemes include the 8B/10B method used by IBM in its ESCON fiber-optic interconnection system for large computers. Most high-speed networks use 4B/5B or 8B/10B encoding schemes. The consequence is that the transmission speed is faster than the data rate.

DATA RATE AND SIGNAL RATE

A close inspection of the modulation codes in Figure 10-4 shows an important aspect of signal transmission. The highs and lows may be changing faster than the 1s and 0s of binary data. In the figure, there are 12 data bits (000100110111). The modulation codes, however, use more symbols to represent these bits. A symbol is a high or a low pulse in the modulation code. The Manchester code uses two symbols, both a high and a low, for each binary bit. It always requires twice as many symbols as there are bits to be transmitted.

When we describe the speed of a system, it becomes important to distinguish between data rate and signal rate. *Data rate* is the number of *data bits* transmitted in bits per second. A system may operate at 10 Mbps. The *signal* rate (or speed) is the number of *symbols* transmitted per second. Signal speed is expressed in baud. The signal speed (or baud rate) and the data rate (or bit rate) may or may not be the same, depending on the modulation code used.

For NRZ data, which uses one symbol per bit, the rates are the same. A 10 Mbps requires a 10 Mbaud capacity. For Manchester-encoded data, which uses two symbols per bit, the baud rate is twice that of the bit rate. To transmit a 10-Mbps data stream requires a transmission link having a bandwidth of 20 Mbaud. The baud rate, or number of symbols to be transmitted, is the true indication of a system's signaling speed. This fact is true for all transmission systems, fiber optic or otherwise.

DUTY CYCLE

Duty cycle refers to the ratio of high to low symbols in the encoded string. A duty cycle of 0 means that all symbols are low. A duty cycle of 100 means that all symbols are high. A duty cycle of 50 means that there are an equal number of highs and lows. Duty cycle expresses the relationship between peak and average power levels arriving at the receiver. For a duty cycle of 50%, the average power is half the peak power. Above a 50% duty cycle, the average power increases in relation to peak power. Below 50%, it declines. Duty cycle becomes important in the receiver, and we will delay further discussion until later in this chapter.

JITTER

Jitter refers to the deviation of pulses from their ideal position in time. Each pulse ideally will occur during its bit period, neither too soon or too late. Think of the clock as a metronome to which the pulses should time themselves. Each pulse should occur with perfect timing in relation to the clock. Jitter is like a musician that can't quite keep time—the pulse gets out of sync. Jitter effects the pulse width of the signal and can increase the probability that the receiver will make an error in detecting the presence or absence of a pulse. Figure 10-5 shows the idea of jitter.

There are three types of jitter:

- *Duty-cycle* or *pulse-width distortion:* this causes the pulse width to be wider or shorter than ideal.
- *Data-dependent jitter:* this is caused by bandwidth limitations in the components. In an optical fiber, data-dependent jitter is caused by dispersion.
- *Random jitter:* this is caused by random sources, such as thermal or shot noise.

FIGURE 10-5
Jitter

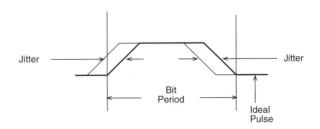

Jitter is the main form of distortion associated with the transmitter and receiver. It can be caused by circuit design, by the electronic components, and by the optical components. Low-jitter operation is mandatory. It should be obvious that jitter control can become more difficult at higher data rates. For a given jitter, the jitter is a lower percentage of a slow data rate than a higher data rate. To return to the metronome example, it is more difficult to keep in time to a fast rhythm. In addition, any mistiming is more likely to move into the beat in front or behind.

EYE PATTERNS

Eye patterns offer a handy qualitative way to evaluate signal forms. An eye pattern is formed by a long stream of random bits superimposed on one another on an oscilloscope. As shown in Figure 10-6, the bit stream includes transitions from high to low, low to high, no transitions, and so forth. You can see both the rising and falling edges on the left and right. The eye pattern shows thousands of such bits displayed simultaneously. By looking at the pattern, you can learn much about the signal quality. In a perfect system, you would only see a series of single lines, since perfection would mean each pulse would fit perfectly over those that came before it. Since the world isn't perfect, the eye pattern will show noise, jitter, and other effects. The center of the pattern is called the *eye*.

The eye pattern can tell you much about the signal.

- The height of the eye measures the noise margin. The higher the eye, the better the noise margin.
- The width of the signals at the center of the eye measures the jitter. The wider the width, the greater the jitter. A thin trace indicates low jitter.
- The height of the signals at the top and bottom measures the noise in the signal. Again, thinner is better.

FIGURE 10-6
Eye pattern

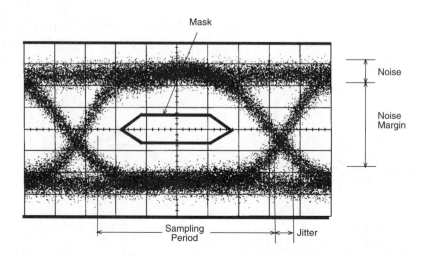

- The rising and falling edges of the pulses show the rise and fall times of the signal.

Often a mask or template is superimposed in the center of the eye. The mask represents the limits of a specific application as far as rise times, fall times, pulse duration, jitter, and noise. As long as the pulses remain outside the mask, the signals are meeting specification. Because eye patterns can provide a great deal of information about the quality of a signal, they are widely used in specifications. Some applications detail each separate specification that can be seen in a pulse, and from which you can build a template for the eye pattern. This includes rise times, fall times, various types of jitter, and so forth. Other applications, notably SONET, use the eye pattern as a basis for specifications.

TRANSMITTER OUTPUT POWER

The output power of the transmitter is of great concern. The power of many transmitters is specified for the power coupled into a given fiber. Thus for a given transmitter design, the typical output power might be as follows:

Fiber	Power
50/125 μm, 0.21 NA	−16 dBm
62.5/125 μm, 0.275 NA	−12 dBm
100/140 μm, 0.30 NA	−8 dBm

The power coupled increases with core diameter and NA. The tradeoff, of course, is that the smaller cores and NA usually indicate fibers with lower attenuation and higher bandwidths.

If the transmitter power output is not specified as listed, the power coupled into the fiber must be calculated. To do so, we must know the source output power, output diameter and NA, fiber core diameter and NA, and the expected connector loss. We will look at an example of this calculation in the next chapter.

BASIC RECEIVER CONCEPTS

Figure 10-7 shows a basic receiver, consisting of detector, preamplifier, quantizer, signal detect circuit, and output buffers. The preamp amplifies the attenuated signal from the detector and converts it from current to voltage. The quantizer accepts the voltage from the preamp, amplifies it further, and converts it to its proper logic state. The quantizer must be able to compensate for noise levels, the affects of variations in temperature, duty cycle, and power supply, and other effects that can make it difficult to properly achieve accurate reception of the incoming data.

FIGURE 10-7
Receiver

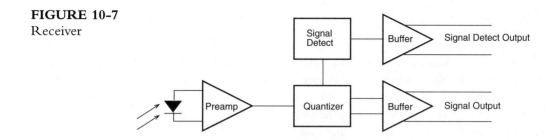

While clock recover—the separation of data and clock signals—is a function of the receiver, the clock-recovery circuit is often not built directly into the receiver package. Instead it is a separate clock-recovery chip or part of an application-specific chip. One reason is that clock recovery tends to be more application-specific. Without clock recovery, a receiver can meet a wider range of applications.

The signal-detect circuit compares the output of the quantizer to a reference signal to determine if the signal-to-noise ratio is sufficient for reliable signal reproduction. If the SNR is not high enough, the signal-detect circuit will change states, in effect sending an alarm that signal levels have fallen below acceptable levels.

The output buffers are line drivers that convert the output to the proper levels for the circuits to which they are connected. For example, the buffer can provide an ECL or TTL output.

Receiver Sensitivity

Receiver sensitivity specifies the weakest optical signal that can be received. The minimum signal that can be received depends on the noise floor of the receiver front end. The last chapter described the limitations on the minimum optical power the detector can receive. The minimum level for the receiver is basically the same, except that the influence of the amplifier noise must also be included. From the noise floor, the lowest *practical* optical power to the receiver depends on the SNR or BER requirements of the system.

Sensitivity can be expressed in microwatts or dBm. Expressing a sensitivity as 1 µW or −30 dBm is saying the same thing. With a packaged receiver, the sensitivity may be specified as an absolute minimum (the noise floor) or with respect to a given level of performance such as BER level.

Dynamic Range

Dynamic range is the difference between the minimum and maximum acceptable power levels. The minimum level is set by the sensitivity and is limited by the detector. The maximum level is set by either the detector or the amplifier. Power levels above the maximum saturate the receiver or distort the signal. The received optical power must be maintained below this maximum.

If a receiver has a minimum optical power requirement of −30 dBm and a maximum requirement of −10 dBm, its dynamic range is 20 dB. It can receive optical inputs between 1 and 100 μW.

Figure 10-8 shows the relationship between BER, data rate, and received power for a typical receiver. As we would expect, BER improves as the received power becomes larger. On the other hand, the required power for a given BER increases significantly as the data rate becomes higher. A BER of 10^{-9} requires an optical input of about −35 dBm (310 nW) for a 25-Mbps data rate and about −32 dBm (620 nW) for a 60-Mbps rate. The received power must be 3 dB, or about 50%, greater for the higher data rate.

AMPLIFIER

Two classical designs used in fiber-optic receivers use a low-impedance amplifier and a transimpedance amplifier. Examples are shown in Figure 10-9.

The bandwidth of the low-impedance amplifier is determined by the RC time constant of the circuit:

$$BW = \frac{1}{2\pi RC}$$

FIGURE 10-8
Relation between received power, BER, and data rate for a typical optical receiver

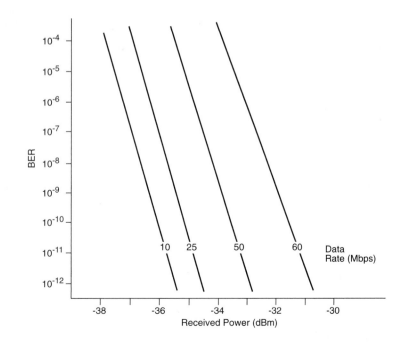

FIGURE 10-9
Low-impedance
and transimped-
ance circuit
designs

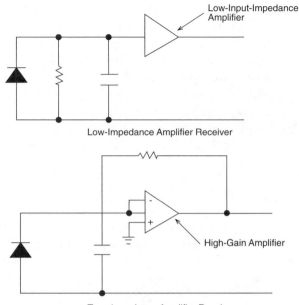

Transimpedance Amplifier Receiver

This equation is the same as for a photodiode whose response time is limited by the time constant. From the discussion of noise and response time for a photodiode, recall that bandwidth and resistor noise decrease with increasing resistance. Therefore, receiver sensitivity can be increased by increasing the load resistor but at the expense of bandwidth. The resulting high-impedance amplifier has a much smaller bandwidth but greater sensitivity.

The bandwidth of a receiver with a transimpedance amplifier is affected by the gain of the amplifier:

$$BW = \frac{g}{2\pi RC}$$

where g is the open-loop gain. To achieve a bandwidth comparable to the low-impedance circuit, the transimpedance circuit must have a much lower value for the resistor across the amplifier.

Representative sensitivities for the different receiver designs operating at 100 Mbps and a BER of 10^{-9} are

- low impedance:
  −33 dBm
- high impedance:
  −40 dBm
- transimpedance:
  −36 dBm

DUTY CYCLE IN THE RECEIVER

The reason for the concern about duty cycle in the modulation codes is that some receiver designs put restrictions on the duty cycle. A receiver distinguishes between high and low pulses by maintaining a reference threshold level. A signal level above the threshold is seen as a high or 1; a signal level below the threshold is seen as a low or 0.

In a restricted-duty-circle receiver, a duty cycle far removed from 50% will increase the chances of a level being misinterpreted. A high level is seen as a low level, or a low level is misinterpreted as a high level. As the duty cycle departs from the 50% ideal, the misinterpretations of signal levels degrade the BER.

Figure 10-10 shows why such errors occur in a receiver with a restricted duty cycle receiving NRZ code. The receiver sets the threshold level based on the average optical power being received. The top part of the figure shows how the threshold level depends on the duty cycle. For a 50% duty cycle, the threshold is set midway between the high and low signal levels, since the average power is midway between signal levels. The difference between the threshold and a low signal level is the same as the difference between the threshold and the high signal level.

At a 20% duty cycle, where low levels significantly outnumber high levels, the average received power is much lower. The threshold shifts much closer to the low signal level. At an 80% duty cycle, the average received power is much higher, and the threshold shifts closer to the high signal level. This shifting results from the average power being received.

This shifting of threshold level would cause no problems in an ideal, noiseless receiver. But receivers are neither perfect nor noiseless. Signal levels not only vary somewhat, but the signals contain noise. The bottom half of the figure shows how noise distorts the pulse. If the signal level crosses the threshold at the wrong point, it is misinterpreted as the wrong level. A 0 can cross the low threshold at a 20% duty cycle and be seen as a 1. A 1

FIGURE 10-10
Errors and duty cycle in a restricted-duty-cycle receiver

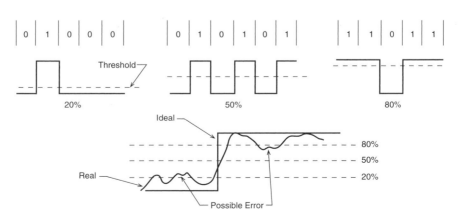

can cross the threshold at an 80% duty cycle and be seen as a 0. In either case, a bit error results.

There are two ways to get around such errors. The first is to maintain a duty cycle close to 50%. Manchester and biphase-M codes, by definition, always have a 50% duty cycle, so they satisfy the requirement. Their drawback is that they require a channel bandwidth of twice the data rate. They also increase the complexity of the transmitter somewhat.

The second method of avoiding bit errors is to design a receiver that maintains the threshold without drift. The reference threshold is always midway between high and low signal levels. One way to do this is by a dc-coupled receiver, which is designed to operate with arbitrary data streams. The receiver is edge-sensing, meaning that it is sensitive to changes in level and not to the levels themselves. This type of receiver reacts only to pulse transitions.

Since the dc-coupled receiver is easy to design and build, why not use it all the time? The tradeoff is receiver sensitivity. It often requires 6 to 8 dB higher optical power levels. This power penalty must be paid by lower data rates, shorter transmission distances, or lower error requirements. Even so, it may often be the best approach in many applications.

In ac coupling, which is one approach used in restricted-duty-cycle receivers, any dc levels are removed from the signal. Only the time-varying portions are left. This stripping of dc levels is often done capacitively, since a capacitor blocks dc and passes ac. The receivers in Figure 10-9 are shown dc coupled to keep the representation simple. An ac-coupled design would place a capacitor between the photodetector and the amplifier.

TRANSCEIVERS AND REPEATERS

Transceivers and repeaters are two important components in many fiber-optic applications. They are simple enough in concept, and we need only comment briefly on each. A *transceiver* is a transmitter and receiver packaged together to allow both transmission and reception from either station. A regenerative *repeater* is a receiver driving a transmitter. The repeater is used to boost signals when the transmission distance is so great that the signal will be too highly attenuated before it reaches the receiver. The repeater accepts the signal, amplifies and reshapes it, and sends it on its way by feeding the rebuilt signal to a transmitter.

One advantage of digital transmission is that it uses regenerative repeaters that not only amplify a signal but reshape it to its original form as well. Any pulse distortions from dispersion or other causes are removed. Analog signals use nonregenerative repeaters, which amplify the signal, including any noise or distortion. Analog signals cannot be easily reshaped, because the repeater does not know what the original signal looked like. For a digital signal, it does know.

As you will see later, an important trend is to use optical amplifiers instead of regenerative repeaters. Optical amplifiers eliminate the need to convert the optical signal to an electronic one and back.

TRANSCEIVER PACKAGING

While you can design your own transmitters and receivers and mount them on a pc board, most are supplied already packaged. They contain all the circuitry needed to accept the data at the transmitter and provide an identical output from the receiver. While transmitters and receivers are available as separate packages, they are more commonly supplied as a transceiver, which contains both the transmitter and receiver in a single package. The rationale for this should be clear. Most applications require communication to and from a piece of equipment. Further, most applications require transmitters and receivers that meet certain requirements. For example, Ethernet specifies the output levels of the transmitter and sensitivity for the receiver. An Ethernet transceiver packs both circuits into a single package. Transceivers can offer economies: The equipment manufacturer needs to design the system for, assemble, and stock only one product.

A trend in transceivers, like most electronics, is to offer smaller and smaller packages. Early transceivers were relatively large 22-pin devices that took up considerable real estate on a pc board. This, in turn, increased the size of the equipment. During the late 80s and early 90s, smaller 9-pin devices that offered about half the depth became popular. The width of the transceiver was constrained by industry-standard connectors like the SC or ST. More recently, the small-form-factor transceiver offered an even smaller size. As discussed in Chapter 11, newer connector interfaces like the MT-RJ offer port densities equal to that of modular jacks. The small-form-factor transceiver accommodates the newer miniature connectors like the MT-RJ and LC. The transceivers still take up more space on the printed circuit board than copper connectors. The ideal for network applications is to create a transceiver that can use the same footprint as a modular jack. This would allow the equipment manufacturer to use the same board design for either copper or fiber ports. Figure 10-11 shows some representative transceivers.

TRANSMITTER AND RECEIVER SPECIFICATIONS

Many of the characteristics for specifying a transmitter or receiver are those of concern with any electronic circuit. These include power supply voltages, storage and operating temperature ranges, required input and output voltage levels (which indicate TTL or ECL compatibility), propagation delays, and so forth. There are also important specifica-

FIGURE 10-11
Transceivers with
MT-RJ and SC
interfaces
(Courtesy of
Hewlett-Packard)

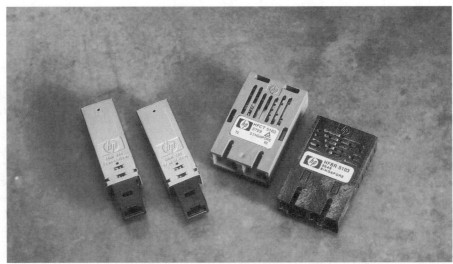

tions relating to the transmitter or receiver in a fiber-optic system. We have already become acquainted with each of these.

Transmitter
- Peak output power
- Data rate/bandwidth
- Operating wavelength
- Source spectral width
- Duty-cycle restrictions
- Rise and fall times
- Jitter

Receiver
- Data rate/bandwidth
- Sensitivity
- Dynamic range
- Operating wavelength
- Duty-cycle restrictions
- Jitter

Figure 10-12 shows specifications for a 155-Mbps ATM transceiver for use with multimode fiber.

FIGURE 10-12
Specifications for an ATM 155-Mbps compatible transceiver

Center Wavelength	1270 to 1380 nm
Transmitter Spectral Width (FWHM)	200 nm
Transmitter Average Power	−20 to −14 dBm
Source Extinction Ratio	10% max
Receiver Sensitivity	−29 to −14 dBm
Rise time/Fall time	0.6 ns min, 2.5 ns max
Duty-Cycle Distortion	0.5 ns max (transmitter)
	0.5 ns max (receiver)
Data-Dependent Jitter	0.5 ns max (transmitter)
	1.0 ns max (receiver)
Random Jitter	0.5 ns max (transmitter)
	0.5 ns max (receiver)

SUMMARY

- A transmitter includes the source and its drive circuit.
- A receiver includes the detector, amplification circuits, and output circuits.
- Modulation codes are the formats by which digital bits are encoded for transmission.
- For some modulation codes, data rate and baud rate are the same. For others, they are not the same.
- Some modulation codes are self-clocking.
- Duty cycle expresses the ratio between high levels and low levels.
- The receiver sensitivity sets the lowest acceptable optical power that can be received.
- The dynamic range of the receiver is the difference between the minimum and maximum optical power that a receiver can handle.

Review Questions

1. Sketch a block diagram of a simple transmitter.
2. A Manchester-encoded transmission has a data rate of 200 Mbps. What is its baud rate?
3. Name two examples of self-clocking modulation codes.
4. Sketch an NRZ pattern for 1111001. Be sure to show each bit period.
5. Will a typical transmitter couple more power into a 50/125-μm fiber or an 85/125-μm fiber? Why?
6. A receiver has a sensitivity of −30 dBm and a maximum receivable power of −10 dBm. What is its dynamic range?
7. Name two of the commonest types of receiver circuits.
8. Sketch a block diagram of a repeater.
9. If a receiver has a sensitivity of −28 dBm, what is the minimum optical power it can receive?

chapter eleven

CONNECTORS AND SPLICES

Interconnection of the various components of a fiber-optic system is a vital part of system performance. Connection by splices and connectors couples light from one component to another with as little loss of optical power as possible. Throughout a link, a fiber must be connected to sources, detectors, and other fibers. This chapter describes the basic considerations involved in fiber-optic interconnections and describes several examples of actual connectors and splices. To simplify the discussion, we emphasize connecting one fiber to another.

By popular usage, a *connector* is a disconnectable device used to connect a fiber to a source, detector, or another fiber. It is designed to be easily connected and disconnected many times. A *splice* is a device used to connect one fiber to another permanently. Even so, some vendors do offer disconnectable splices that are not permanent and that can be disconnected for repairs or rearrangement of circuits.

The requirements for a fiber-optic connection and a wire connection are very different. Two copper conductors can be joined directly by solder or by connectors that have been crimped or soldered to the wires. The purpose is to create intimate contact between the mated halves to maintain a low-resistance path across the junction. Connectors are simple, easy to attach, reliable, and essentially free of loss.

The key to a fiber-optic interconnection is precise alignment of the mated fiber cores (or mode field diameter in single-mode fibers) so that nearly all the light is coupled from one fiber across the junction into the other fiber. Contact between the fibers is not even mandatory. The demands of precise alignment on small fibers create a challenge to the designer of the connector or splice.

THE NEED FOR CONNECTORS AND SPLICES

Connectors and splices are required for many reasons. Although these reasons may be fairly obvious since they mirror the same needs with metallic conductors, we will briefly review them here.

In a long link, fibers may have to be spliced end to end because cable manufacturers offer cables in limited lengths—typically 1 to 6 km. If the cable comes in 6-km lengths, a 30-km span requires four splices (as well as connections to the transmitter and receiver). In other cases, it may not be practical to even pull a 6-km length of cable through ducts during installation. More moderate lengths may be easier to install.

Connectors or splices may be required at building entrances, wiring closets, couplers, and other intermediate points between transmitter and receiver. These allow, for example, transitions between outdoor and indoor cables, rearrangement of circuits, and the division of optical power from one fiber into several fibers.

Finally, fibers must be connected to the source in the transmitter and the detector in the receivers.

Dividing a fiber-optic system into several subsystems connected together by splices and connectors also simplifies system selection, installation, and maintenance. Components may be selected from different vendors. They may be installed by different vendors or contractors. For example, a building contractor can wire a building with fiber-optic cable, a manufacturer of computer terminals can build fiber-optic transmitters and receivers in the equipment, and the telephone company can bring a fiber-optic telephone line into the building. All these various parts are then linked together with connectors. Maintenance of the system is simplified when a faulty or outdated part can be disconnected and a new or updated part installed. Faster transmitters and receivers, for instance, can be installed without disturbing the fiber.

CONNECTOR REQUIREMENTS

The following is a list of desirable features for a fiber-optic connector or splice:

- *Low loss:* The connector or splice should cause little loss of optical power across the junction.
- *Easy installation:* The connector or splice should be easily and rapidly installed without need for extensive special tools or training.
- *Repeatability:* A connector should be able to be connected and disconnected many times without changes in loss.
- *Consistency:* There should be no variation in loss; loss should be consistent whenever a connector is applied to a fiber.

■ *Economical:* The connector or splice should be inexpensive, both in itself and in special application tooling.

It can be very difficult to design a connector that meets all the requirements. A low-loss connector may be more expensive than a high–loss connector, or it may require relatively expensive application tooling. The lowest losses are desirable, but the other factors clearly influence the selection of the connector or splice.

In general, loss requirements for splices and connectors are as follows:

■ 0.2 dB or less for telecommunication and other splices in long-haul systems.
■ 0.3 to .75 dB for connectors used in intrabuilding systems, such as local area networks or automated factories.
■ 1 to 3 dB for connectors and splices used in applications where such higher losses are acceptable and low cost becomes more important than low loss. Such applications often use plastic fibers.

CAUSES OF LOSS IN AN INTERCONNECTION

Three different types of factors cause loss in fiber-optic interconnections:

1. Intrinsic or fiber-related factors are those caused by variations in the fiber itself.
2. Extrinsic or connector-related factors are those contributed by the connector itself.
3. System factors are those contributed by the system.

INTRINSIC FACTORS

In joining two fibers to each other, we would like to assume that the two fibers are identical. Usually, however, they are not. The fiber manufacturing process allows fibers to be made only within certain tolerances. From the nominal specification, a fiber will vary within stated limits. Figure 11-1 shows schematically the most important variations in tolerances that cause loss. The following paragraphs discuss them in greater detail.

NA-mismatch loss occurs when the NA of the transmitting fiber is larger than that of the receiving fiber. *Core-diameter-mismatch loss* occurs when the core or diameter of the transmitting fiber is larger than that of the receiving fiber. *Cladding-diameter-mismatch loss* occurs when the claddings of the two fibers differ, since the cores will no longer align.

FIGURE 11-1
Factors of intrinsic interconnection loss

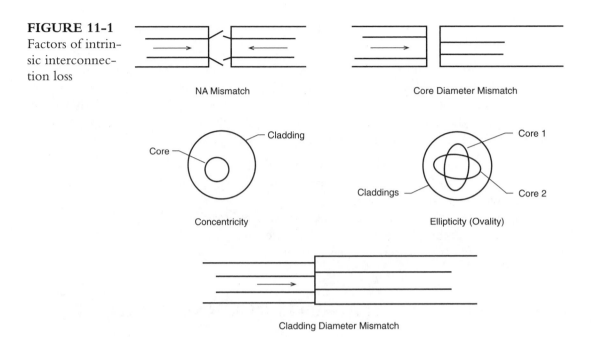

Concentricity loss occurs because the core may not be perfectly centered in the cladding. Ideally, the geometric axes of the core and cladding should coincide. The concentricity tolerance is the distance between the core center and the cladding center.

Ellipticity (or *ovality*) *loss* occurs because the core or cladding may be elliptical rather than circular. Notice that the alignment of the two elliptical cores will vary, depending on how the fibers are brought together. At one joining, they may be at the extremes of ellipticity so that maximum loss occurs. At the next joining, one fiber may become rotated in relation to the other, so their elliptical axes are the same. No loss then occurs. The ellipticity or ovality tolerance of the core and cladding equals the minimum diameter divided by the maximum diameter.

These variations exist in any fiber. The fiber manufacturer controls these variations by manufacturing fiber to tight, exacting tolerances. In the past few years, manufacturing techniques have improved sufficiently that fiber tolerances are much tighter. For example, a 125-μm-diameter fiber once had a tolerance of 0.5 μm, so the actual fiber diameter was in a range of 120 to 130 μm. Two fibers at the extreme of tolerance mismatch present a loss of 0.6 dB. Today, the normal tolerance is 0.2 μm, resulting in a range of 123 to 127 μm and a maximum loss of 0.28 dB. A tolerance of 0.1 μm reduces the loss to 0.1 dB.

Remember that these losses are maximums and will not be experienced in most cases since the probability of joining two fibers at the extreme diameter mismatches is small.

EXTRINSIC FACTORS

Connectors and splices also contribute loss to the joint. When two fibers are not perfectly aligned on their center axes, loss occurs even if there is no intrinsic variation in the fibers. The loss results from the difficulty of manufacturing a device to the exacting tolerances required. As we will see, many different alignment mechanisms have been devised for joining two fibers.

The four main causes of loss that a connector or splice must control are

1. Lateral displacement
2. End separation
3. Angular misalignment
4. Surface roughness

The following discussion assumes perfect fibers, without tolerance variation. The curves presented are general ones for multimode fibers. As discussed later in this chapter, modal conditions in the fiber also influence loss across the junction.

Lateral Displacement

A connector should align the fibers on their center axes. When one fiber's axis does not coincide with that of the other, loss occurs. As shown in the graph in Figure 11-2, the amount of displacement depends on the ratio of the lateral offset to the fiber diameter. Thus the acceptable offset becomes less as the fiber diameter becomes smaller. A displacement of 10% of the core axis yields a loss of about 0.5 dB. For a fiber with a 50-μm core, a 10% displacement is 5 μm, which, in turn, means that each connector half must contribute only a 2.5 μm.

FIGURE 11-2
Loss from lateral displacement (Illustration courtesy of AMP Incorporated)

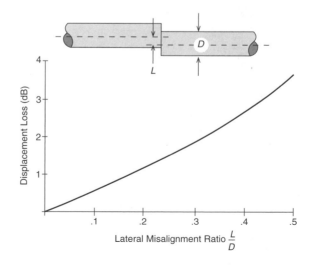

These dimensions would double for a fiber with a 100-μm core: a 10% displacement would allow 10 μm or 5 μm for each connector half. Obviously, then, control of lateral displacement becomes more difficult for smaller-diameter fibers. Connector manufacturers attempt to limit displacement to less than 5% of the core diameter.

End Separation

Two fibers separated by a small gap experience two types of loss, as shown in Figure 11-3. The first is *Fresnel reflection loss,* which is caused by the difference in refractive indices of the two fibers and the intervening gap, usually air. Fresnel reflection occurs at both the exit from the first fiber and at the entrance to the second fiber. For glass fibers separated by air, loss from Fresnel reflections is about 0.34 dB. Fresnel losses can be greatly reduced by using an index-matching fluid between the fibers. *Index-matching fluid* is an optically transparent liquid or gel having a refractive index the same as or very near that of the fibers.

The second type of loss for multimode fibers results from the failure of high-order modes to cross the gap and enter the core of the second fiber. High-order rays exiting the first fiber will miss the acceptance cone of the second fiber. Light exiting the first fiber spreads out conically. The degree of separation loss, then, depends on fiber NA. A fiber with a high NA cannot tolerate as much separation to maintain the same loss as can a fiber with a lower NA.

Ideally, the fibers should butt to eliminate loss from end separation. In most splices, the fibers do butt. In a separable connector, a very small gap is sometimes desirable to prevent the ends from rubbing together and causing abrasion during mating. Fibers brought together with too much force, as may be done by over-tightening coupling nuts on a connector, may even be fractured. Therefore, some connectors are designed to maintain a very

FIGURE 11-3

Loss from end separation (Illustration courtesy of AMP Incorporated)

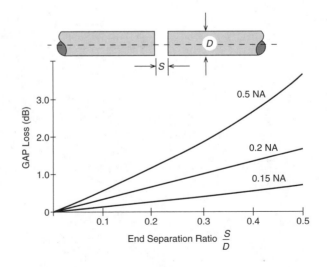

small gap between fibers, and others use spring pressure to bring the fiber ends gently together without the danger of damage. Physical contact of the fibers is often desirable to control return reflections (which are discussed later in this chapter) and end gap losses.

Angular Misalignment

The ends of mated fibers should be perpendicular to the fiber axes and perpendicular to each other during engagement. The losses graphed in Figure 11-4 result when one fiber is cocked in relation to the other. Again, the degree of loss also depends on the fiber NA. Notice, however, that the influence of NA is opposite that for end separation. Here larger NA can tolerate greater angular misalignment to maintain the same loss as for a lower NA.

A properly applied connector controls angular misalignment rather easily, so its loss contribution is not as great as with lateral displacement. This control includes having the fiber face perpendicular to the fiber axis, which is a function of proper fiber cleaving or polishing, and preventing the connector from cocking during engagement.

Surface Finish

The fiber face must be smooth and free of defects such as hackles, burrs, and fractures. Irregularities from a rough surface disrupt the geometrical patterns of light rays and deflect them so they will not enter the second fiber.

SYSTEM-RELATED FACTORS

The loss at a fiber-to-fiber joint depends not only on the losses contributed by the connector and fiber but on system-related factors as well. Chapter 6 discussed how modal

FIGURE 11-4

Loss from angu-
lar misalignment
(Illustration
courtesy of AMP
Incorporated)

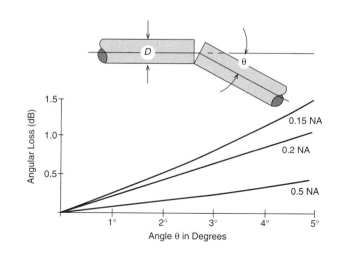

conditions in a fiber change with length until the fiber reaches equilibrium mode distribution (EMD). Initially, a fiber may be overfilled or fully filled, with light being carried both in the cladding and in high-order modes. Over distance, these modes will be stripped away. At EMD, a graded-index fiber has a reduced NA and a reduced active area of the core carrying the light.

Consider a connection close to the source. The fiber on the transmitting side of the connection may be overfilled. Much of the light in the cladding and high-order modes will not enter the second fiber, although it was present at the junction. This same light, however, would not have been present in the fiber at EMD, so it would also not have been lost at the interconnection point.

Next consider the receiving side of the fiber. Some of the light will spill over the junction into cladding and high-order modes. If we were to measure the power from a short length of fiber, these modes would still be present. But these modes will be lost over distance, so their presence is misleading.

Similar effects will be seen if the connection point is far from the source, where the fiber has reached EMD. Since the active area of a graded-index fiber has been reduced, lateral misalignment will not affect loss as much, particularly if the receiving fiber is short. Again, light will couple into cladding and high-order modes. These modes will be lost in a long receiving fiber.

Thus, the performance of a connector depends on modal conditions and the connector's position in the system (since modal conditions vary along the length of the fiber). In evaluating a fiber-optic connector or splice, we must know conditions on both the launch (transmitter) side and the receive (receiver) side of the connection. Four different conditions exist:

1. Short launch, short receive
2. Short launch, long receive
3. Long launch, short receive
4. Long launch, long receive

With all other things held constant, connector performance depends on the launch and receive conditions. For a series of tests done on one connector, for example, losses under long-launch conditions were in the 0.4- to 0.5-dB range. Under short-launch conditions, the losses were in the 1.3- to 1.4-dB range. Thus, a difference of nearly 1 dB occurred simply as a result of the launch conditions used in the test.

INSERTION LOSS

Insertion loss is the method for specifying the performance of a connector or splice. Power through a length of fiber is measured. Next, the fiber is cut in the center and the connector or splice is applied. Power at the end of the fiber is again measured. Insertion loss is

$$\text{loss}_{\text{IL}} = 10 \log_{10}\left(\frac{P_1}{P_2}\right)$$

where P_2 is the initial measured power and P_1 is the measured power after the connector is applied.

This testing method, by using a single fiber, minimizes the influence of fiber variation on loss. By rejoining the cut fiber, we make the mated halves essentially identical. The only fiber variations present are eccentricity and ellipticity. NA and diameter variations are eliminated. The purpose is to evaluate connector performance independently of fiber-related variations.

Launch and receive conditions still affect the measured insertion loss. Therefore, these conditions must be known when comparing insertion loss specifications for various connectors. Differences in loss specifications may result from differences in test conditions. In addition, the application in which the connector will be used must also be considered. It is best to evaluate a connector that has been tested under conditions similar to those in which it will be used.

ADDITIONAL LOSSES
IN AN INTERCONNECTION

The discussion of intrinsic fiber loss stated that loss occurs when the transmitting fiber has a core diameter or NA greater than the core diameter or NA of the receiving fiber. This discussion assumed the connection was between two fibers of the same type, differing only in variations in tolerance. In this case, these variations may affect the performance of the connector, but the losses from the mismatches are not usually calculated in general loss budgets. If, however, two different types of fibers are connected, diameter- and NA-mismatch losses may be significant and must be accounted for.

When the NA of the transmitting fiber is greater than that of the receiving fiber, the NA-mismatch loss is

$$\text{loss}_{\text{NA}} = 10 \log_{10}\left(\frac{\text{NA}_r}{\text{NA}_t}\right)^2$$

Similarly, when the diameter of the transmitting fiber is greater than that of the receiving fiber, the diameter-mismatch loss is

$$\text{loss}_{\text{dia}} = 10 \log_{10}\left(\frac{\text{dia}_r}{\text{dia}_t}\right)^2$$

Assume, for example, a transmitting fiber with a core diameter of 62.5 μm and an NA of 0.275 and a receiving fiber with a diameter of 50 μm and an NA of 0.20. Such a connection could be experienced in connecting fibers in a local area network. Loss from NA-mismatch is

$$
\begin{aligned}
\text{loss}_{NA} &= 10 \log_{10}\left(\frac{NA_r}{NA_t}\right)^2 \\
&= 10 \log_{10}\left(\frac{0.20}{0.275}\right)^2 \\
&= -2.8 \text{ dB}
\end{aligned}
$$

Loss from diameter mismatch is

$$
\begin{aligned}
\text{loss}_{dia} &= 10 \log_{10}\left(\frac{dia_r}{dia_t}\right)^2 \\
&= 10 \log_{10}\left(\frac{50}{62.5}\right)^2 \\
&= -2.9 \text{ dB}
\end{aligned}
$$

The total loss is 5.7 dB.

LOSS IN SINGLE-MODE FIBERS

Connectors and splices for single-mode fibers must also provide a high degree of alignment. In many cases, the percentage of misalignment permitted for a single-mode connection is greater than for its multimode counterpart. Because of the small size of the fiber core, however, the actual dimensional tolerances for the connector or splice remain as tight or tighter. For example, a gap of 10 times the core diameter results in a loss of 0.4 dB in a single-mode fiber.

Mismatches in mode field diameter (spot size) are the greatest cause of intrinsic, fiber-related loss. Most manufacturers limit spot size deviations to 10%, which results in attenuation increases of under 0.1 dB in 99% of the cases where fibers from the same manufacturer are connected. Since different manufacturers have different values for mode field diameter, greater loss may be expected when mating connectors from different manufacturers.

RETURN REFLECTION LOSS

You learned earlier that when two fibers are separated by an air gap, optical energy will be reflected back toward the source. This energy is termed *return reflection* or *return loss*. The loss refers to the fact that it is energy representing loss at the fiber-to-fiber juncture. It is that part of the loss that reflects and is propagated in a backwards direction by the fiber. Figure 11-5 shows the idea of return reflections.

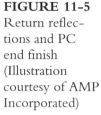

FIGURE 11-5
Return reflec-
tions and PC
end finish
(Illustration
courtesy of AMP
Incorporated)

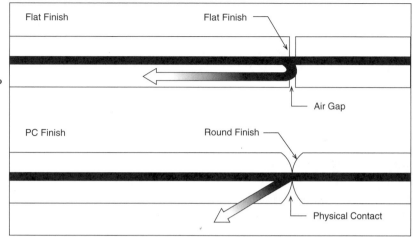

In a single-mode interconnection with flat end finish, this loss amounts to −11 dB—that is, the reflected energy is about 11 dB below the incident energy. If 500 μW of energy reach the fiber end, about 40 μW will reflect back toward the source. This level of energy is sufficient to interfere with operation of a laser diode. In single-mode systems, it is especially important to minimize return loss so that proper operation of the source is not disrupted.

Either by assuring fiber-to-fiber contact or eliminating air from the fiber interface, return loss can be lowered to over −30 dB—the 500 μW now reflect only 0.5 μW toward the source. One way to do this is by rounding the fiber end during polishing. Interconnected fibers can now be allowed to touch, which curtails reflection from the mismatches in refractive indices of air and fibers. Only minor reflections occur because of small differences in fiber properties.

Why not use a flat end finish and butt the fibers? Because achieving two *perfectly* flat, *perfectly* perpendicular ends is difficult. Most likely, one or both of the fibers will have a slight angle, enough to keep the fiber cores from touching. With a rounded finish, fibers always touch on the high side near the light-carrying fiber core.

A rounded PC end finish can find application in multimode systems as well. In some local area networks, for example, the system relies on dead silence between packets of information. The silence indicates that there are no transmissions on the network and that it is OK for a station to transmit. If not reduced enough, return reflections echoing through the network can be interpreted as a carrier, disrupting operation.

Another approach to increasing return reflection loss is to angle the fiber end, which serves to reflect light in the cladding. The angle is slight—8° or 9°.

Several variations of PC end finishes have been devised to give different levels of return loss:

PC: >30 dB
SuperPC: >40 dB
UltraPC: >50 dB

The important thing is that the geometry and finish of the ferrule and fiber are essential to achieving high return loss.

FIBER TERMINATION

A connector or splice is used to terminate the fiber. In most cases, a splice is simpler than a connector. The permanence of splices allows a connection device that has fewer component parts. A fusion splice, in fact, has no parts in its simplest form. A fiber-optic connector or splice must fulfill several functions:

- Align the fibers optically
- Secure the fiber in the connector or splice
- "Decouple" the fiber from the cable. By this, we mean that the cable, usually the strength members, is also secured so that a pull or tension on the cable remains on the cable rather than on the more fragile fiber.

Figure 11-6 shows a cross-sectional view of two mated connectors. The connectors are held aligned with each other by a coupling receptacle, which is basically a precision-toleranced bore with an outside that provides a means to fasten the connector. The receptacle is typically either metal or plastic with a resilient metal, ceramic, or plastic sleeve to align the connector.

Ferrules

Most connectors use a ferrule to hold the fiber and provide alignment. In one sense, many connectors are simply different ways to build a practical connector around a ferrule.

Ceramic ferrules offer the best performance and are the preferred material for single-mode fibers. Ceramic ferrules are strong and have precision, machined fiber bores. In addition, they have excellent thermal and mechanical properties so that performance does not vary due to temperature or environmental fluctuations.

FIGURE 11-6
Mated connectors (Illustration courtesy of AMP Incorporated)

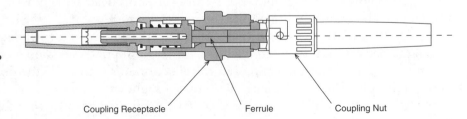

Coupling Receptacle   Ferrule   Coupling Nut

Plastic ferrules offer lower cost but more modest performance. Stainless steel ferrules fall between ceramic and plastic in performance and cost, but they are popular because of their strength—they are less susceptible to breakage or cracking than ceramic.

Two types of ceramic are used for ferrules: alumina oxide and zirconia oxide. Alumina oxide was the first material used: It is a hard, inelastic material that allows manufacturers to hold tolerances very precisely. Its coefficient of thermal expansion—how much the material expands when heated and contracts when cooled—is very close to that of glass. The drawback to alumina is that it is more brittle and can break under excess stress. In addition, alumina is much harder to polish, especially in a field installation. Alumina ferrules are seldom used today.

Zirconia oxide is softer than alumina oxide and more resistant to impact. It is still hard enough to hold tolerances nearly as well as alumina oxide, but soft enough to allow faster polishing. Zirconia ferrules are the most popular today.

The most popular ferrule size is a 2.5-mm diameter, which has become nearly a standard, although newer small-form-factor connectors use a 1.25-mm ferrule.

Epoxy and Polish

Fibers are most often epoxied into the connector. Epoxy, because of its curing time, is generally considered to be an undesirable but necessary step in a fiber-optic termination. Epoxy provides good tensile strength to the termination to prevent the fiber from moving within the connector body. It also prevents pistoning—the in/out movement of the fiber within the ferrule. One cause of pistoning is changes in temperature. With changes in temperature, the fiber can expand out of the ferrule or contract within it. If the fiber pistons outward, it extends past the end of the ferrule and can be damaged. If the fiber pistons inward, it increases the gap between fibers and thereby the loss. After the epoxy cures, the fiber/connector is polished to a smooth end finish.

Epoxyless Terminations

Even so, epoxyless connectors have been devised. One type uses an internal insert that grips the fiber at the front and rear, providing the stability and tensile strength of epoxy. As the connector is crimped, the insert compresses around the fiber at both the front and the rear.

The front clamp is on the bare fiber. This clamp prevents pistoning. It does not, however, provide the tensile strength to completely prevent movement of the fiber in the connector, and increasing the clamping pressure at this point runs the risk of damaging the fiber. The rear compression adds higher clamping force on the fiber buffer coating to provide the necessary tensile strength. This gripping point will not prevent pistoning during changes in temperature. It is only when both compressions are properly balanced that the fiber is securely held within the connector enough that epoxy is not needed.

Another approach is to install a small stub of fiber in the ferrule at the factory. A cleaved fiber is inserted and clamped into place, forming a low-loss splice with the stub. The advantage of this approach is that it eliminates the need for polishing. The fiber/ferrule are finished at the factory and do not require further field polishing.

Performance is comparable to its epoxy counterparts. The main advantage of an epoxyless connector is speed of assembly. Some users will tolerate a slightly higher loss to achieve a fast, easy termination. Time, after all, is money. A new, large building being wired for fiber may contain thousands of connectors. Speed of installation may be more important than achieving losses much lower than required by the application.

Notice that the epoxyless approach is a technique that is not limited to one connector style.

COMPATIBILITY

A growing trend in electronics is standardization. Standardization implies that a component—anything from a connector to a computer—meets specified requirements and is interchangeable with other components meeting that standard. The two components are compatible. Whereas some standards are *de facto* ones—they evolve through popularity and not formal standards—most standards are formally created by such industry associations as the Electronics Industry of America (EIA), the Institute of Electrical and Electronic Engineers (IEEE), or the American National Standards Institute (ANSI).

Compatibility here refers to the need for the connector to be compatible with other connectors or with specifications. For example, large major telecommunication companies have developed connectors for use in their equipment and applications. For many applications, it becomes important for a competitive connector to mate with these.

Compatibility exists on several levels. The most basic level is physical compatibility: The connector must meet certain dimensional requirements to allow it to mate with other connectors of the same style.

The next level of compatibility involves performance: insertion loss, durability, temperature range, and so forth.

Finally, standards like MIL-C-83522 define the connector completely: dimensions, performance, materials. Here, compatibility comes to mean identical since the military specification leaves very little room for differences.

CONNECTOR EXAMPLES

The following pages describe some of the most common connectors in use today.

The late 1980s and early 1990s saw great progress in standardization of connectors. Bear in mind that compatibility does not mean connectors are identical. The great drive in compatibility is the interface: the ability of connectors from different vendors to mate with one another. This still leaves room for different manufacturers to differentiate their products. For example, the original FC connector had fifteen separate parts that had to be assembled by the technician. Some vendors today offer FC-style connectors with only four separate parts. Such design improvements make the connector easier to use.

Most single-mode connectors use ceramic ferrules. Although a single-mode connector can be used with multimode fibers, a multimode connector should not be used with single-mode fibers. The reason is tolerances. A connector for 125-μm *multimode* fiber must have a ferrule bore large enough to accommodate the largest fiber size of 127 μm. A connector for *single-mode* fiber is often available for the specific tolerance size; that is, you can obtain the connector with a ferrule bore diameter of 125, 126, or 127 μm to fit the fiber exactly. This also means that you may have several connectors on hand to try on the fiber until you get a snug fit. Having a connector with a 126-μm ferrule bore and a fiber with a 127-μm diameter doesn't work.

FC-style Connector

The earliest connector to be based on the 2.5-mm ceramic ferrule, the FC connector was originally devised by Nippon Telegraph and Telephone for telecommunications (Figure 11-7). It has been very popular in Japan, Europe, and the U.S. MCI, for example, used it in its fiber-optic telephone network in the 1980s.

FIGURE 11-7
FC-style and D4-style connectors (Photo courtesy of AMP Incorporated)

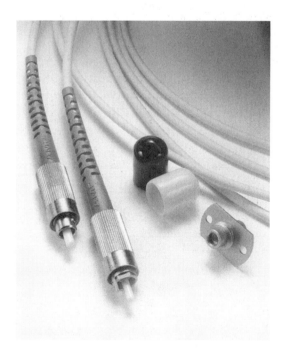

The connector uses a threaded coupling nut, which has the advantage of providing a secure connection even in high-vibration environments and the disadvantage of not permitting quick connection or disconnection. The coupling nut must be rotated several times to thread or unthread the connector.

The connector also offers tunable keying. A keyed connector with a small key results in the ferrule always mating in its adapter bushing the same way. It cannot rotate between one mating and the next; this minimizes any changes due to variations in concentricity or ellipticity in ferrule or fiber (see Figure 11-1). Tunable keying means that the key can be adjusted to the point of lowest loss. For example, insertion loss measurements can be made with the key in several positions. The key is then locked into position.

Although tuning is useful in applications where the lowest possible loss is required, it is not used in most applications because of the small difference of 0.1 or 0.2 dB gained by tuning.

The connector is available in both single-mode and multimode versions. The earliest version of the FC used a flat ferrule endface, but the newer FC/PC uses a rounded PC endface to permit higher return loss. Some manufacturers offer connectors with an angled endface to achieve the same result.

D4-style Connector

This connector is very similar to the FC connector, with its threaded coupling, tunable keying, and PC end finish. The main difference is its 2.0-mm-diameter ferrule. It was originally designed by the Nippon Electric Corporation (Figure 11-7).

ST-style Connector

The ST connector, which was designed by AT&T Bell Laboratories for use in premises wiring of buildings and other applications, uses the same 2.5-mm ceramic ferrule as the FC connector but with a quick-release bayonet coupling (Figure 11-8). Quick-release couplings are preferred in applications where severe vibrations are not expected—such as offices.

With over thirty manufacturers offering ST connectors, it is probably the most popular connector style. It is widely used in local area networks, premises wiring, test equipment, and many other applications. Many applications that specify other connectors allow the ST to be used as an alternative.

The quick-release bayonet locking mechanism requires only a quarter turn during mating or unmating, while built-in keying ensures repeatable performance during mating. The key ensures that the fiber is always inserted to the mating bushing with the same orientation: One fiber will not be rotated with respect to another. Often, predictable, consistent loss is more important than lowest loss.

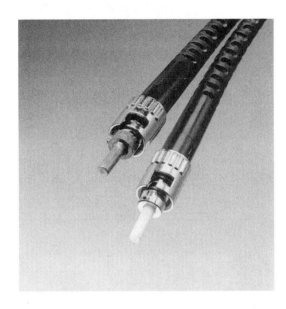

FIGURE 11-8
ST-style con-
nectors (Photo
courtesy of
AMP
Incorporated)

Because of its popularity, the ST connector is offered in several variations: with ceramic, stainless steel, or plastic ferrule; in single-mode and multimode versions; and in epoxyless versions. Insertion loss is around 0.3 dB.

SC Connectors

The SC connector achieved wide popularity in the early 1990s for both single-mode and multimode applications (Figure 11-9). Originally designed by Nippon Telegraph and Telephone, the connectors use a push-pull engagement for mating. The SC is designed to be pull-proof. In a pull-proof connector, the ferrule is decoupled from the cable and connector housing. A slight pull on the cable will not pull the ferrule out of optical contact with the interconnection.

The name SC comes from "subscriber connector," which describes its original application in telecommunications. It is replacing the FC and D4 connectors in new telecommunications applications worldwide, and it is a strong competitor to the ST connector in local area networks, premises wiring, and similar applications.

The basic SC connector consists of a plug assembly containing the ferrule. The plugs mate into connector housings. One attractive feature of the SC is the ease with which multifiber connectors are constructed, with multiple-position housings or with clips that hold two or more plugs together. Connectors that require twisting, like the FC or ST, are not adaptable to multifiber applications.

FIGURE 11-9
SC connectors
(Photo courtesy
of AMP
Incorporated)

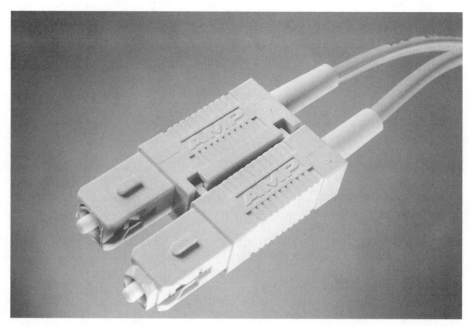

An important application of this multifiber capability is duplex applications, where one fiber carries information in one direction and the other fiber carries information in the other direction. For example, a computer workstation with a fiber-optic connection to the network contains both a transmitter and a receiver. A duplex connector to both permits a single connector to plug into both the transmitter and receiver.

MU Connector

The MU connector is a miniaturized version of the SC. It uses a 125-mm ferrule and is about one-quarter the size of an SC.

FDDI MIC Connector

Designed by ANSI for use in FDDI networks (described in detail in Chapter 15), this connector is a duplex connector using two 2.5-mm ferrules (Figure 11-10). The distinguishing feature is the fixed shroud that protects the ferrules from damage. A floating interface ensures consistent mating without stubbing.

A positive side-latch mechanism and keying capability (per FDDI) make the connector easy to use. Companion coupling adapters allow an FDDI connector to mate with another FDDI connector, with two ST-style connectors, or with a transceiver. The term

FIGURE 11-10
FDDI MIC
connector

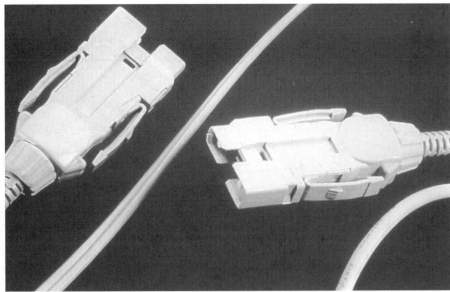

MIC means *medium interface connector,* referring to the connector that serves the interface between electronics and fiber transmission medium.

Because of their features—duplex configuration, low loss, wide availability, and easy use—these connectors are popular beyond FDDI applications.

ESCON Connector

This connector's name derives from its application in IBM's ESCON channel interface mentioned in Chapter 1 and described in Chapter 15 (Figure 11-11). It is similar to the FDDI connector—a duplex connector using 2.5-mm ferrules and a floating interface. The principal difference is its retractable shroud, which pulls back during engagement of the connector with a transceiver. While this simplifies the design of the transceiver interface, it may not provide as much protection as a fixed shroud.

Plastic–Fiber Connectors

Connectors exclusively for plastic fibers stress *very* low cost and easy application—often with no polishing or epoxy. Plastic fiber can be trimmed to an acceptable finish with a hot knife (similar to a cross between an X-acto knife and a soldering iron), obviating the need for polishing. Fiber retention is mechanical, usually by gripping the fiber cladding with barbs or other secure means. This allows very fast application, with much less skill than required for acceptable terminations with glass fibers.

FIGURE 11-11
ESCON connectors (Photo courtesy of AMP Incorporated)

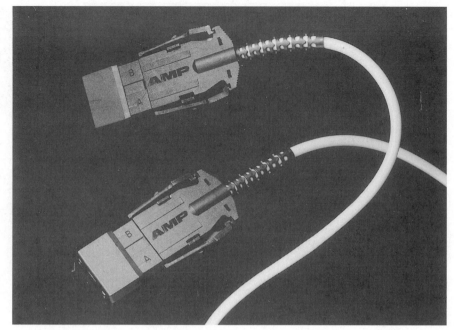

Plastic fiber connectors include both proprietary designs and standard designs such as the digital audio connector described in Electronics Industry of Japan (EIAJ) RC-5720. Connectors such as the ST or SMA are also available for use with plastic fibers. As plastic-fiber technology gains popularity in such applications as digital audio electronics and other consumer applications, there will undoubtedly be greater standardization. In other applications—automotive or security systems, for instance—standardization is less critical.

Small-Form-Factor Connectors

Connectors such as the FC, ST, and SC were the standard connectors for most applications throughout the late 1980s and early 1990s. However, compared to electrical connectors they were large and less convenient to use. While they are clearly embedded and will continue to find use, a new generation of connectors offer smaller size and easier use. These are sometimes called small-form-factor connectors and are considerably smaller than ST and SC connectors.

One measure of size is found in network applications. A common piece of network equipment mounts in a 19-inch rack. How many connectors you can fit into a given space is sometimes called the port density. It is easy to fit 24 copper RJ-45 connectors in a line within this 19-inch width. With a duplex SC connector, you can fit only 12. Thus, RJ-45 connectors offer twice the port density of SC connectors. The new generation of small-form-factor connectors allow the same 24 ports, yielding a port density equal to copper. This simplifies design for network equipment manufacturers and saves space in equipment closets. Chapter 15 will explore network and premises cabling applications further. For now, note that small-form-factor connectors overcome some of the drawbacks in size, cost, and convenience that have hampered the wider use of fiber optics in buildings.

MT-Style Array Connectors

An array connector has more than one fiber held within the ferrule. The most popular style is the MT connector invented by NTT in the 1980s for use in telephone equipment in Japan. During the 1990s, connectors based on the MT ferrule and its derivatives have become popular for numerous other applications.

The MT ferrule holds up to 12 single-mode or multimode fibers in an area smaller than the single-fiber SC connector. Most MT-style connectors use a push-pull mechanism to maintain two plug connectors mated in an adapter. The MT ferrule has been adopted for different styles of connectors. The most popular is the industry-standard MPO connector, which is used for wiring data centers and other applications.

Although the connectors can be used with a variety of fiber constrictions, ribbon fiber is very popular. Figure 11-12 shows an MPO connector, along with an SC connector for comparison. The SC connector holds one fiber; the MPO holds 12 fibers in a connector significantly smaller that the SC. This size reduction is one reason why MT-style and other small-form-factor connectors are gaining in popularity.

FIGURE 11-12 12-fiber MPO connector (right), shown in comparison to a single-fiber SC connector (Photo courtesy of AMP Incorporated)

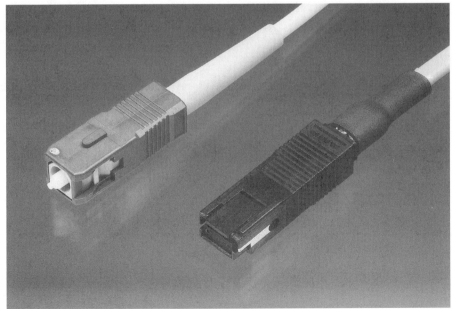

One other variation of the MT-style is the MPX, which is also a multifiber connector holding up to 12 fibers. The MPX is widely used inside equipment. It can connect to array transceivers or be used as a backplane connector to carry signals off of a board.

For convenience, we will refer to connectors having two fibers as *duplex* connectors and connectors capable of holding more than two fibers as *multifiber* or *array* connectors. As will become more apparent in Chapter 15, where we will explore some applications, array connectors are most often used behind the walls, under the floor, and inside equipment. They are a behind-the-wall application in most cases. The main exception is two-fiber connectors, such as the MT-RJ and LC, which are used for patch cables and other equipment connections.

MT-RJ Connector

The MT-RJ connector was designed by AMP, Siecor, and USConec as an alternative to RJ-45 modular telephone connectors. Modular connectors are arguably the most widely used and best known connectors in the world. Just look at your telephone. It uses 4- or 6-position modular plugs and jacks. RJ-45 connectors are the 8-position copper connectors used in network and building cable applications. The MT-RJ uses a ferrule smaller than the standard MT to hold two fibers. It also uses a press-to-release latch quite similar to that found on the modular jack to make its operation familiar. Figure 11-13 shows an MT-RJ connector, along with the small-form-factor transceiver it uses.

Most fiber-optic connectors use two plug connectors mated through a coupling adapter. While the MT-RJ can be used in this manner, it also offers a true plug-jack interconnection. Designed for very simple termination of fiber, the jack contains two fiber stubs in a polished ferrule. The backend of the stub fibers are in a mechanical splice. To terminate the fiber-optic cable, the fibers are cleaved, slid into the splice, and the splice is closed to form a low-loss interconnection.

FIGURE 11-13
MT-RJ connector and companion transceiver (Photo courtesy of AMP Incorporated)

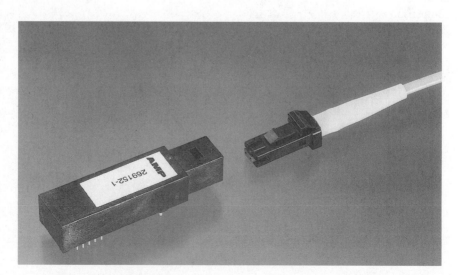

LC Connector

The LC connector was designed by Lucent Technologies as a smaller connector for premises cabling, data networking, and telecommunications applications. Today the LC is made by several companies. Figure 11-14 shows the connector. Based on a 125-mm ceramic ferrule, the connector uses an RJ-45-style housing with a push-to-release latch that provides an audible click when the connector is mated in an adapter. Available in both duplex and simplex configurations, the connectors double the density over existing ST and SC designs. The connectors are color-coded beige for multimode and blue for single mode.

Volition Connector

The Volition VF-45 connector from 3M is a ferruleless connector, shown in Figure 11-15. The plug connector consists of a fiber holder to secure the fiber, a shroud and boot that secures the connector to the cable, and a hinged door that acts as a dust cover. The mating socket has a fiber holder, a hinged door, and a body that houses V-grooves to align the fibers. When the plug is inserted into the socket, the fibers slide along the V-grooves until they mate with the fiber in the socket (also in the V-groove). The fibers in the plug bend, creating a forward and downward pressure to optimize the fiber-to-fiber contact.

FIGURE 11-14
LC small-form-factor connector manufactured by Methode Electronics (Courtesy of Methode Electronics)

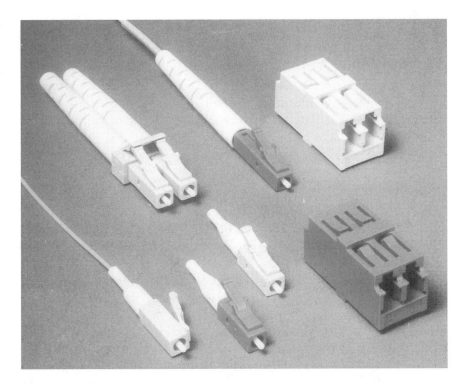

FIGURE 11-15
Volition VF-45
connector
(Courtesy of
3M)

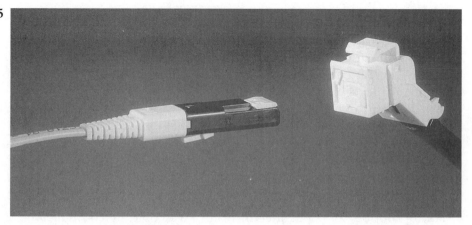

The plug snaps into place when fully inserted and uses a press-to-release latch similar to modular telephone plugs.

The Volition connector has been adopted as a standard connector in Fibre Channel applications. You can see a trend here: the new generation of connectors are designed to be as small as—and as easy to use as—modular plugs and jacks.

SPLICES

Fusion Splice

Fusion splicers use an electric arc to weld two fibers together. Fusion splicers offer very sophisticated, computer-controlled alignment of fibers to achieve the lowest losses routinely achieved—as low as 0.05 dB. Because they fuse the fiber, they virtually eliminate return reflections. Their main drawback is the high cost of equipment. Nevertheless, fusion splices remain the choice where the lowest possible losses are required.

Mechanical Splices

Several forms of mechanical splices have been devised. These all share common elements: They are easily applied in the field, require simple or no tooling, and offer losses on the order of 0.1 to 0.2 dB. Some splices are reenterable—a telephone industry term meaning that they can be reused.

While the main movement in connectors is toward offering such industry-standard designs as the SC, MT-RJ, or LC connector, each splice design remains proprietary to the manufacturer. The reason is simple: They don't have to be compatible because they do not mate with anything.

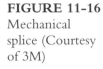

FIGURE 11-16
Mechanical splice (Courtesy of 3M)

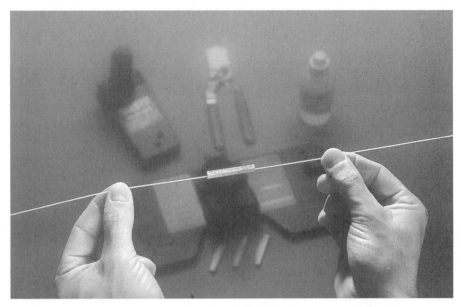

FIBER PREPARATION

Proper preparation of the fiber end face is critical to any fiber-optic connection. The two main features to be checked for proper preparation in end finish are *perpendicularity* and *end finish*. The end face ideally should be perfectly square to the fiber and practically should be within 1° or 2° of perpendicular. Any divergence beyond 2° increases losses unacceptably. The fiber face should have a smooth, mirrorlike finish free of blemishes, hackles, burrs, and other defects.

The two common methods used to produce correct end finishes are the *scribe-and-break method* (also called the score-and-break method) and the *polish method*. The scribe-and-break method is used mostly with splices, whereas the polish method is more commonly used with connectors.

Whichever method is used, the fibers must first be bared. Jackets, buffer tubes, and other outer layers are removed with wire strippers, cutting pliers, utility knives, and other common tools. Metal strength members are removed with wire cutters, and scissors are used to cut Kevlar strength members.

The plastic buffer coating attached to the cladding can be removed chemically or mechanically. Chemical removal involves soaking the fiber end for about 2 min in a paint stripper or other solvent and then wiping the fiber clean with a soft tissue. Mechanical removal is done with a high-quality wire stripper. Care must be taken not to nick or damage the cladding. The bared fiber can be cleaned with isopropyl alcohol or some other suitable cleaner.

The scribe-and-break method uses a fiber held under slight pressure. A cutting tool with a hard, sharp blade, such as diamond, sapphire, or tungsten carbide, scribes a small nick across the cladding. The blade can be pulled across a stationary fiber, or a fiber can be pulled across a stationary blade. After scribing, the pressure is increased by pulling, which forces the flaw to propagate across the face of the fiber.

Properly done, the cleave produces a perpendicular, mirrorlike finish. Improperly done, the cleave results in hackles and lips, as shown in Figure 11-17, that make the cleave unacceptable. The process must then be repeated.

Scribe-and-break cleaving can be done by hand or by tools that range from relatively inexpensive hand tools to elaborate automated bench tools. Any technique or tool is capable of good cleaves; the trick is consistent finishes time and time again. In general, the less costly approaches require more skill and training for the technicians making the cleave.

Most connectors use polishing to achieve the proper end finish after the fiber has been partially or completely assembled in the connector. Polishing is the final step of assembly. Most connectors use some sort of polishing fixture, which has a large polishing surface to

FIGURE 11-17
Good and bad cleaves of an optical glass fiber (Photos courtesy of GTE Fiber Optic Products)

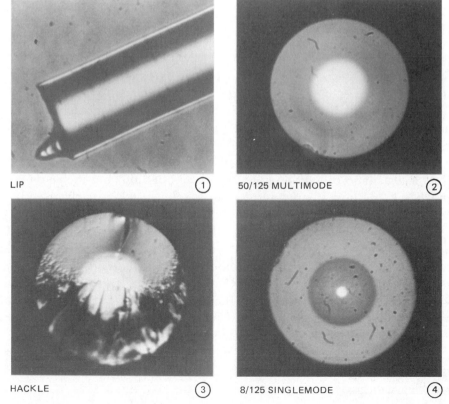

LIP ①   50/125 MULTIMODE ②

HACKLE ③   8/125 SINGLEMODE ④

ensure a perpendicular finish. A smaller-faced fixture runs the risk that the assembly will be cocked or tilted during hand polishing.

Polishing is done in two or more steps with repeatedly finer polishing grits, typically down to 1 μm or 0.3 μm. Polishing is done with a figure-8 motion. The connector and fiber face should be cleaned before switching to a finer polishing material.

As with a cleaved fiber, the polished fiber should be inspected under a microscope. Small scratches on the fiber face are usually acceptable, as are small pits on the outside rim of the cladding. Large scratches, pits in the core region, and fractures indicate unacceptable end finishes. Some poor finishes, such as scratches, can be remedied with additional polishing with 1-μm, or finer, film. Fractures and pits usually mean a new connector must be installed. Figure 11-18 shows examples of acceptable and unacceptable polishes.

CONNECTOR ASSEMBLY EXAMPLE

To demonstrate the practical assembly of a connector, we reproduce in this section the assembly instructions for an SC-style connector from AMP Incorporated (Figure 11-19).

FIGURE 11-18 Good and bad polishes (Photos courtesy of Buehler Ltd., 71 Waukegan Road, Lake Bluff, IL 60044)

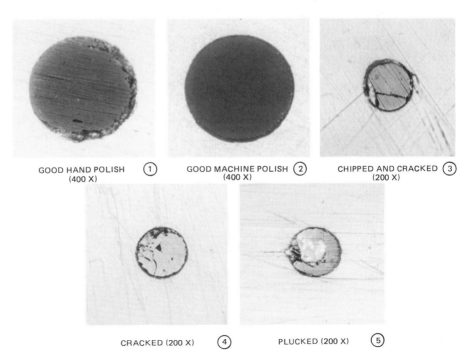

GOOD HAND POLISH ① (400 X)

GOOD MACHINE POLISH ② (400 X)

CHIPPED AND CRACKED ③ (200 X)

CRACKED (200 X) ④

PLUCKED (200 X) ⑤

Instruction Sheet
408–4363
12 DEC 96   Rev O

**Duplex Multimode Polymer
SC Connector Kits**

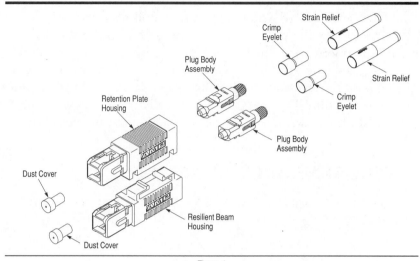

Figure 1

1. INTRODUCTION

This instruction sheet covers the application of AMP*
Duplex Multimode Polymer SC Connector Kits, such
as 492209–[ ], to fiber optic cable.

Read this material thoroughly before starting
assembly.

 NOTE *Dimensions on this sheet are in millimeters [with
inch equivalents in brackets]. Figures and
illustrations are for identification only, and are not
drawn to scale.*

2. DESCRIPTION (Figure 1)

The connector kit consists of two crimp eyelets, strain
reliefs, plug bodies, connector housings, and dust
covers. This connector is to be used with 125–μm
optical fiber with a 500– or 900–μm buffer in a 3 mm
[0.12 in.], 2.5/2.4 mm [0.10/0.09 in.], or 2 mm
[0.08 in.] jacket, where the buffer has movement
relative to the jacket.

3. ASSEMBLY PROCEDURE

3.1. Required Tools and Materials

The following AMP tools and materials are required
for applying the Duplex Multimode Polymer SC

Connector Kit to optical fibers (numbers in
parentheses are corresponding AMP instruction
sheets):

A. Termination Tools and Consumables

— Cable Stripper 501198–1 (408–9394)
— Crimp Tool 58588–1 *(tool only)*
— 3 mm [0.12 in.] Die Assembly 503910–1
(die assembly only)
— 3 mm [0.12 in.], 2.5/2.4 mm [0.10/0.09 in.], and
2 mm [0.08 in.] Die Assembly 492025–1
(die assembly only)
— Crimp Tool with 3 mm [0.12 in.] Die Assembly
503911–1 *(tool and die assembly)*
— Crimp Tool with 3 mm [0.12 in.], 2.5/2.4 mm
[0.10/0.09 in.], and 2 mm [0.08 in.] Die
Assembly 492026–1 *(tool and die assembly)*
— Epoxy 502418–1 or 501195–4
— Epoxy Applicator Kit 501473–3
— Epoxy Mixer 501202–1
— 203 μm [0.008 in.] Fiber Stripper 504024–1
(408–9485)
— Fiber Protector 502656–1
— Scissors 501014–1
— Cleave Tool 504064–1
— Cable Preparation Template Kit 492215–1
— Curing Oven 502134–1 (110V),
502134–2 (240V) (408–9460)

1 of 7

LOC B

FIGURE 11–19
Connector assembly example

Duplex Multimode Polymer SC Connector Kits   408–4363

B. Hand Polishing Tools and Consumables

— Polishing Bushing 502631–1
— Polishing Plate 501197–2
— Polishing Pad 501858–1 (Green Pad)
— 5–μm Polishing Film 228433–8 (pkg. of 10)
— 1–μm Polishing Film 228433–7 (pkg. of 10)
— 0.3–μm Polishing Film 228433–5 (pkg. of 10)
— Cotton Swabs
— Pipe Cleaner

C. Machine Polishing Tools and Consumables

ITEM	MACHINES		
	SFP–550	SFP–100D	SFP–70D
Machine w/ SC Holder	503556–2	503730–2	503559–2
SC Holder Only	503558–2	503732–2	503561–2
SC Spacer (ea., order 10)	504669–2	504669–2	504669–2

— Microscope Kit 502970–1 (408–9801)
— Alcohol Pads or Isopropyl Alcohol
— Distilled Water
— Double–Sided Tape
— Lint–Free Tissues or Cloths

Specific instructions for machine polishing are provided with the polishing machine.

3.2. Preparing Fibers

DANGER *Always wear eye protection when working with optical fibers. Never look into the end of terminated or unterminated fibers. Laser radiation is invisible but can damage eye tissue. Never eat, drink, or smoke when working with fibers. This could lead to ingestion of glass particles.*

DANGER *Be very careful to dispose of fiber ends properly. The fibers create slivers that can easily puncture the skin and cause irritation.*

1. Cut the cable 25 mm [1 in.] longer than the required finish length.

2. Slide strain relief onto the cable; then slide on the crimp eyelet with the larger diameter toward the end of the cable. See Figure 2.

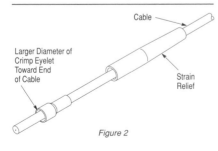

Figure 2

3. Strip the cable to the dimensions shown in Figure 3 using Cable Stripper 501198–1, Fiber Stripper 504024–2, and Scissors 501014–1. Use the AMP Template 492215–1 for recommended strip dimensions and tolerances.

NOTE: Not to Scale. Dimensions are in Inches

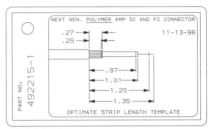

Figure 3

4. Clean the fiber thoroughly using an alcohol pad.

CAUTION *Never clean a fiber with a dry tissue.*

5. Slide the plug assembly over the stripped fiber to ensure the proper size and strip length. The fiber should protrude slightly past the end of the connector and the buffer should stop against the back of the ferrule before the jacket stops against the connector end.

3.3. Selecting and Preparing Epoxy

A. Selecting Epoxy

Epoxy 501195–4 is easy to work with because it comes in packs with premeasured components. It will cure in 24 hours at 25°C [77°F]; or using Epoxy Curing Oven 502134–1, in 2 hours at 65°C [150°F] or 10 minutes at 110°C [230°F].

Epoxy 502418–1 is bulk packaged in two tubes. Using Epoxy Curing Oven 502134–1, it will cure in 30 minutes at 100°C [212°F].

B. Preparing Epoxy 501195–4

1. Remove the separating clip from the epoxy package and mix the epoxy thoroughly for 20 to 30 seconds. use of AMP Epoxy Mixer 501202–1 is recommended for thorough mixing of the components.

2. Install the needle tip on the Epoxy Applicator 501473–3. Make sure it is secure. Remove the plunger.

FIGURE 11–19
continued

194 FIBER-OPTIC COMPONENTS

3. Cut the epoxy packet open and squeeze the epoxy into the back of the applicator. Replace the plunger. Hold the applicator vertically and slowly push on the plunger until the entrapped air escapes and a bead of epoxy appears at the tip.

NOTE *An alternative method in using the epoxy is to remove the tip of the epoxy applicator by twisting it one–quarter turn and pulling it away from the body of the applicator. Put the open end of applicator into epoxy and pull back on plunger to draw epoxy into the applicator. See Figure 4. Put tip on the applicator; make sure it is firmly secured. Hold the applicator vertically and slowly push on plunger until the entrapped air escapes and a bead of epoxy appears at the tip.*

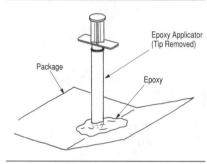

Figure 4

C. Preparing Epoxy 502418–1

1. Squeeze equal lines of the two components onto a clean, disposable surface.

2. Mix thoroughly with a wooden stick.

3. Install the needle tip on the epoxy applicator. Make sure it is secure. Remove the plunger.

4. Load the epoxy into the back of the applicator. Replace the plunger. Hold the applicator vertically (needle upward) and slowly push on the plunger until all the entrapped air escapes and a bead of epoxy appears at the tip.

4. TERMINATING FIBERS

1. Hold connector in a vertical position. Clean applicator tip and insert it into the tube recessed in the connector end until it stops. See Figure 5.

CAUTION *The applicator tip must be clean.*

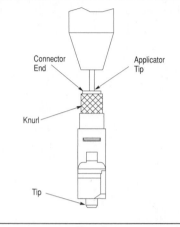

Figure 5

NOTE *Do not get epoxy on the outside of the tube.*

2. Slowly inject epoxy through the applicator until the epoxy appears at the tip of the connector.

3. Stop injecting epoxy and retract the applicator slightly (.76 mm [.030 in.]), and pause for one second.

4. Withdraw the applicator quickly, without injecting additional epoxy into the connector.

CAUTION *If too much epoxy is injected into the connector, it will not function properly.*

5. Carefully insert the fiber and buffer into the connector while flaring the strength members evenly over the knurl of the connector. See Figure 5. The fiber should appear at the tip. The buffer should stop against the back of the ferrule before the jacket stops against the connector end. Do not allow the strength members to enter the connector end. See Figure 5.

5. CRIMPING

Die assembly must be installed correctly in tool frame.

1. Squeeze the handles on the hand tool until the ratchet releases. Open the tool fully.

2. Position crimp eyelet over the strength members which surround the knurled portion of the connector.

FIGURE 11-19
continued

AMP Duplex Multimode Polymer SC Connector Kits 408–4363

3. Place the large diameter of the crimp eyelet into the large diameter of the die. Align the edge of the large diameter of the crimp eyelet with the edge of the crimp die. When using Die Assembly 492026–1, be sure to use the proper die position for the cable diameter. See Figure 6.

4. Squeeze the crimp tool handles shut to crimp large diameter of the eyelet.

NOTE *If using Die Assembly 503911–1 for 3 mm [0.12 in.] cable only, proceed to Step 5.*

If using Die Assembly 492026–1, proceed to Step 7.

Die Assembly 503911–1

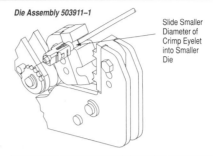

Slide Smaller Diameter of Crimp Eyelet into Smaller Die

Figure 7

Die Assembly 503911–1

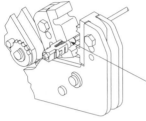

Place Large Diameter of Crimp Eyelet into Large Diameter of Die

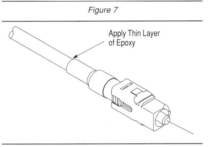

Apply Thin Layer of Epoxy

Figure 8

Die Assembly 492026–1

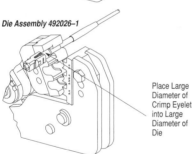

Place Large Diameter of Crimp Eyelet into Large Diameter of Die

9. Slide connector housing over the plug body until it clips in place. The chamfer should be aligned with the housing as shown in Figure 9. In addition, be sure to orient the resilient beam and retention plate housings with the features positioned as shown in Figure 9. This will be critical when joining the two housings to form a duplex connector.

NOTE *If you plan to machine polish, do not complete this step until after polishing is complete.*

Figure 6

5. Slide the smaller diameter of the crimp eyelet into the smaller die until it stops. See Figure 7.

6. Squeeze the crimp tool handles shut to crimp small diameter of the eyelet.

7. Apply a thin layer of epoxy to the crimp eyelet/cable jacket interface. See Figure 8.

8. Slide the strain relief toward the connector until it covers the entire crimp eyelet.

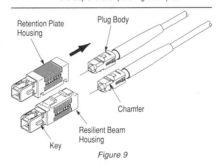

Retention Plate Housing

Plug Body

Chamfer

Resilient Beam Housing

Key

Figure 9

FIGURE 11-19
continued

AMP Duplex Multimode Polymer SC Connector Kits 408–4363

10. Install a fiber protector onto the connector housing. See Figure 10.

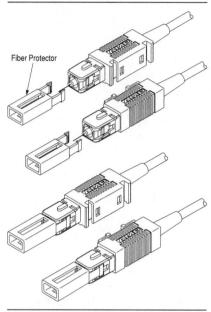

Fiber Protector

Figure 10

11. Position the connector vertically with the ferrule down and cure as follows:

For Epoxy 501195–4
Oven Cure: 2 hours at 65µC [150µF]
or 10 minutes at 110µC [230µF]
Ambient Cure: 24 hours at 25µC [77µF]

For Epoxy 502418–1
Oven Cure: 30 minutes at 100µC [212µF]

12. Remove any epoxy from the crimping dies using an alcohol pad.

6. CLEAVING

1. Remove the fiber protector from the connector.

2. Firmly supporting the connector assembly, use Cleave Tool 504064–1 to scribe the fiber. Draw the beveled edge of the tool across the fiber as shown in Figure 11. After scoring the fiber, pull fiber straight away from the connector to finish the cleave process.

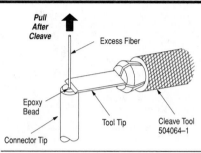

Pull
After
Cleave

Excess Fiber

Epoxy Bead

Tool Tip

Cleave Tool 504064–1

Connector Tip

Figure 11

DANGER *Safely dispose of excess fiber.*

CAUTION *Do not allow the cleave tool to make contact with the epoxy. This may damage the tool tip.*

3. Gently push on the ferrule to ensure tip movement. If the ferrule does not move in an axial direction, too much epoxy was used and the cable must be reterminated.

7. POLISHING

7.1. Hand Polishing Procedures

1. To level off the fiber, lightly polish the end face with hand–held 5–µm polishing film.

2. Install the connector into the Polishing Bushing 502631–1. See Figure 12.

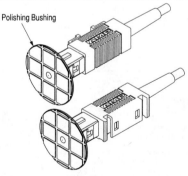

Polishing Bushing

Figure 12

FIGURE 11-19
continued

AMP   Duplex Multimode Polymer SC Connector Kits   408–4363

3. Place the Polishing Pad 501858–1 on the Polishing Plate 501197–2. The polishing pad and plate are used throughout the remainder of the polishing procedure.

4. Place the 5–μm polishing film on the polishing pad.

CAUTION *Always place the polishing bushing on a clean area of the polishing film. Never start polishing on or across a dirty section of the film.*

5. Holding the polishing bushing and connector, start polishing very lightly. Polish in an elongated figure–8 pattern. See Figure 13. Initially, a small amount of exposed fiber will be worn away. This is indicated by a narrow white trace on the film. As the exposed fiber wears away, the trace will widen and darken, indicating that epoxy is being removed. At this point, a slight downward force may be applied while polishing. Check the tip often and stop polishing on the 5–μm film when the epoxy is medium blue and about one–third the ferrule diameter in size.

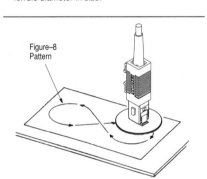

Figure–8 Pattern

Figure 13

6. Clean the polishing bushing and connector assembly with an alcohol pad.

7. Remove the 5–μm film from the polishing pad and replace it with the 1–μm film.

NOTE *Examine the tips frequently to avoid over–polishing, for example, every two figure–8 patterns.*

CAUTION *When using the 5–μm film, it is essential that not all of the epoxy is removed. Polishing must stop when the epoxy changes to light blue.*

8. Using a light force, polish in figure–8 pattern on the 1–μm film until all epoxy is removed.

9. If a superior polish is required, repeat Step 8 using .3–μm Polishing Film 228433–5. Apply only a light force to the bushing.

NOTE *For a smoother end face, use a drop or two of water during polishing.*

10. Remove the connector from the polishing bushing and clean it with isopropyl alcohol and wipe dry with a lint–free cloth or tissue.

11. To inspect the polished fiber, see Section 9, INSPECTING THE FIBER.

7.2. Machine Polishing Procedures

The connectors can be polished on the SFP–550, 100D, and 70D polishing machines. The procedures are identical for each machine. See the table on page 2 for part numbers for each machine. Machine operating instructions are provided with the machine.

8. FINAL ASSEMBLY

NOTE *Once joined, the resilient beam and retention plate housings are NOT separable.*

1. The retention plate housing will retain the resilient beam housing when the two are joined. Join the two housings by holding the resilient beam housing and sliding the retention plate housing over the beams until they latch in place. See Figure 14.

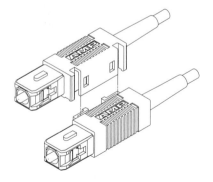

Figure 14

FIGURE 11–19
continued

AMP   Duplex Multimode Polymer SC Connector Kits   408–4363

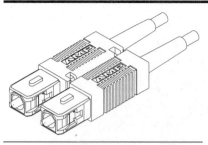

Figure 15

2. Figure 15 shows a completely assembled duplex multimode polymer SC connector.

9. INSPECTING THE FIBER

DANGER   *Never inspect or look into the end of a fiber when optical power is applied to the fiber. The infrared light used, although it cannot be seen, can cause injury to the eyes.*

NOTE   *Refer to AMP instruction sheet 408–9801 for information on the use of the AMP Microscope Kit 502970–1.*

- Be sure all epoxy is removed from the ferrule.
- Dirt may be mistaken for small pits. If dirt is evident, clean with an alcohol pad and then dry.
- Large scratches on the fiber end face indicate too much pressure was used when polishing on the 5–μm film or the connector was over–polished on the 0.3–μm film.
- Fine polishing lines are acceptable. See Figure 16.
- Small chips at the outer rim of the fiber are acceptable. Large chips in the center of the fiber render the polish unacceptable, and the fiber must be re–terminated.

10. CLEANING PROCEDURES

DANGER   *Compressed air used for cleaning must be reduced to less than 206.8 kPa [30 psi], and effective chip guarding and personal protective equipment (including eye protection) must be used.*

10.1. One Connector Removed From Receptacle

1. Wipe completely around the connector's ferrule with an alcohol pad two times.

2. Repeat step 1 using a dry lint–free cloth.

3. Place the dry lint–free cloth on a smooth, flat surface. Holding the connector perpendicular, wipe the ferrule across the cloth.

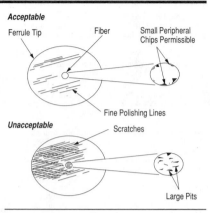

Figure 16

4. Blow compressed air across the end face of the ferrule.

5. Examine the end face for debris, using Microscope Kit 502970–1.

6. Remate the connector to the receptacle.

7. If the optical loss is unacceptable, repeat Steps 1 through 6.

8. If the value is still too high, unmate both connectors and follow procedures in Paragraph 10.2, Two Connectors Removed From Receptacle.

10.2. Two Connectors Removed From Receptacle

1. Wipe completely around both connectors' ferrules with an alcohol pad two times.

2. Repeat step 1 using a dry lint–free cloth.

3. Place the dry lint–free cloth on a smooth, flat surface. Holding the connector perpendicular, wipe both the ferrules across the cloth.

4. Blow compressed air across the end face of both the ferrules.

5. Examine the end faces for debris, using Microscope Kit 502970–1.

6. Blow compressed air through the receptacle.

7. Remate the connectors to the receptacle.

8. If the optical loss is unacceptable, repeat Steps 1 through 7.

9. If the value is still not acceptable, unmate the connector, clean the mating ferrule and alignment sleeve, and repeat Steps 1 through 7.

FIGURE 11–19
continued

SUMMARY

- A splice is used for permanent and semipermanent connections.
- A connector is used for disconnectable connections.
- Loss in an interconnection results from intrinsic factors, extrinsic factors, and system-related factors.
- There are many different designs for connectors and splices. The common ingredient in all designs is precise alignment of fibers.
- Fiber ends can be prepared by the scribe-and-break technique (cleaving) or by polishing.

 Review Questions

1. What is the purpose of a splice or connector?
2. What role does the alignment mechanism play in a fiber–optic connector or splice?
3. Name three sources of intrinsic loss in an interconnection.
4. Name three sources of extrinsic loss in an interconnection.
5. Describe how modal patterns affect loss in a fiber-to-fiber interconnection. Will the loss probably be lower if the transmitting fiber is fully filled or at EMD? Why?
6. Assume an application requiring five 10-km reels of fiber to be joined in a 50-km link. Would connectors or splices be preferred? Why?
7. Would a connector or splice be preferred to connect a fiber-optic cable to a personal computer? Why?
8. Calculate the total mismatch loss for a transmitting fiber with a 62.5/125-μm diameter and a 0.29 NA and a receiving fiber with an 85/125-μm diameter and a 0.26 NA. Ignore other sources of loss for this example.
9. What is the insertion loss test meant to measure? What does the test attempt to eliminate as a loss factor?
10. Name the two critical results a good cleave must achieve.

chapter twelve

COUPLERS, MULTIPLEXERS, AND OTHER DEVICES

Thus far, we have viewed the fiber-optic link as a point-to-point system—one transmitter linked to one receiver over an optical fiber. Even a duplex link is point to point—one transceiver communicates with another transceiver over a fiber-optic pair. In many applications, however, it is desirable or necessary to divide light from one fiber into several fibers or, conversely, to couple light from several fibers into one fiber. In short, we wish to distribute the light. A *coupler* is a device that performs such distribution. Figure 12-1 shows the idea of a coupler: It divides or combines light. The figure also shows an example of an actual device.

This chapter describers couplers and their uses. Important application areas for couplers are in networks, especially local area networks, and in wavelength-division multiplexing (WDM).

COUPLER BASICS

A *coupler* is a multiport device. A *port* is an input or output point for light. There are several types of loss associated with a coupler. Figure 12-2 shows a four-port directional coupler that we will use to define ideas important to your understanding of couplers. Arrows indicate the possible directions of flow for optical power through the coupler. Light injected into port 1 will exit through ports 2 and 3. Ideally, no light will appear at port 4. Similarly, light injected into port 4 will also appear at ports 2 and 3 but not at port 1.

FIGURE 12-1
Couplers (Photo
courtesy of AMP
Incorporated)

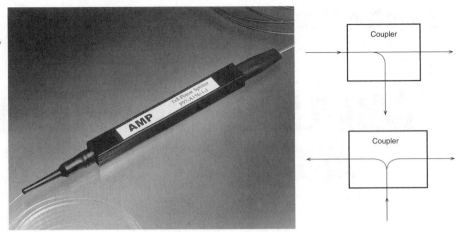

The coupler is passive and bidirectional. Ports 1 and 4 can serve as input ports, ports 2 and 3 as output ports. Reversing the direction of power flow allows ports 2 and 3 to serve as input ports and port 1 and 4 to serve as output ports.

For the following discussion of loss, we assume port 1 is the *input* port and ports 2 and 3 are the *output* ports. Furthermore, the power at port 2 is always equal to or greater than the power at port 3. Therefore, we will term port 2 as the *throughput* port. Port 3 is the *tap* port. These terms are used to suggest that a path containing the greater part of the power is the throughput path, whereas the path containing the lesser part is the tapped path.

Throughput loss is the ratio of output power at port 2 to input power at port 1:

$$\text{loss}_{\text{THP}} = 10 \log_{10} \left(\frac{P_2}{P_1} \right)$$

Tap loss is the ratio of the output power at port 3 to the input power at port 1:

$$\text{loss}_{\text{TAP}} = 10 \log_{10} \left(\frac{P_3}{P_1} \right)$$

Directionality is the ratio between unwanted power at port 4 and the input power at port 1:

$$\text{loss}_{\text{D}} = 10 \log_{10} \left(\frac{P_4}{P_1} \right)$$

FIGURE 12-2
Four-port
directional coupler

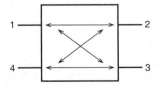

Ideally, no power appears at port 4, so $\text{loss}_D = 0$. Practically, some power does appear through such mechanisms as leakage or reflections. Directionality is sometimes called *isolation*. Directionality or isolation in a good tap is 40 dB or greater; only a very small amount of light appears at port 4.

Excess loss is the ratio between output power at ports 2 and 3 to the input power at port 1:

$$\text{Loss}_E = 10 \log_{10}\left(\frac{P_2 + P_3}{P_1}\right)$$

Excess losses are losses that occur because the coupler is not a perfect device. Losses occur within fibers internal to the coupler from scattering, absorption, reflections, misalignment, and poor isolation. In a perfect coupler, the sum of the output power equals the input power ($P_2 + P_3 = P_1$). In a real coupler, the sum of the output power is always something less than the input power because of excess loss ($P_2 + P_3 < P_1$).

Excess loss does not include losses from connectors attaching fibers to the ports. We looked at these losses and their causes in the last chapter. Furthermore, since most couplers contain an optical fiber at each port, additional loss can occur because of diameter and NA mismatches between the coupler port and the attached fiber.

The input power must obviously be divided between the two output ports. The *coupler splitting ratio* is simply the ratio between the throughput port and the tap port: P_2/P_3. Typical ratios are 1:1, 2:1, 3:1, 6:1, 10:1, 90:10, and 99:1. From the splitting ratio, the relation of the throughput loss and tap loss is constant. Figure 12-3 shows the losses for perfect taps of given splitting ratios.

In a real coupler, the losses at the output ports are the sums of the individual loss and the excess loss. If $\text{Loss}_{THP'}$ and $\text{Loss}_{TAP'}$ represent throughput and tap loss in a real coupler, actual losses become:

$$\text{Loss}_{THP'} = \text{Loss}_{THP'} + \text{Loss}_E$$

$$\text{Loss}_{TAP'} = \text{Loss}_{TAP'} + \text{Loss}_E$$

Assume a directional coupler with a 3:1 splitting ratio and an excess loss of 1 dB. Actual losses become 2.25 dB for throughput power and 8 dB for tap power. If 100 μW of optical power are input to port 1, how much output power appears at each port? Since loss is the ratio of power at the two ports, we know that the throughput power at port 2 equals

$$2.25 \text{ dB} = 10 \log_{10}\left(\frac{P_2}{100 \text{ μW}}\right)$$

The equation, notice, accounts for excess loss. The 2.25 dB includes 1.25 dB throughput loss plus 1 dB for excess loss. We rearrange and solve the equation to find the power at port 2:

FIGURE 12-3
Losses for ideal
four-port direc-
tional couplers

SPLITTING RATIO	THROUGHPUT LOSS (dB)	TAP LOSS (dB)
1:1	3	3
2:1	1.8	4.8
3:1	1.25	6
6:1	0.66	8.5
9:1	0.46	10
10:1	0.41	10.4

$$P_2 = \left[\log_{10}^{-1} \left(\frac{-2.25 \text{ dB}}{10} \right) \right] 100 \text{ } \mu\text{W}$$

$$= 60 \text{ } \mu\text{W}$$

(Notice that since loss is a negative quantity, we make the 2.25 dB negative in the equation.) Tap power at port 3 is

$$P_3 = \left[\log_{10}^{-1} \left(\frac{-7 \text{ dB}}{10} \right) \right] 100 \text{ } \mu\text{W}$$

$$= 20 \text{ } \mu\text{W}$$

A directional coupler is symmetrical, so that the losses remain the same regardless of which ports serve as the input port, throughput port, tap port, and isolated port.

TEE COUPLER

A *tee coupler* is a three-port device. Figure 12-4 shows the application is a typical bus network. A coupler at each node splits off part of the power from the bus and carries it to a transceiver in the attached equipment. If there are many nodes on the bus, the couplers typically have a large splitting ratio so that only a small portion of the light is tapped at each node. The throughput power at each coupler is much greater than the tap power.

Tee couplers are of greatest use when there are a few terminals on the bus. Consider a bus having N terminals. A signal must pass through $N - 1$ couplers before arriving at the

FIGURE 12-4
Tee network

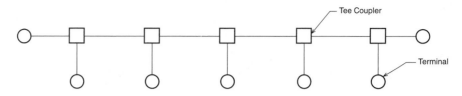

receiver. In a coupler having only throughput and tap losses (i.e., no excess loss), the total distribution loss is

$$L = (N - 1)\text{Loss}_{\text{THP}} + \text{Loss}_{\text{TAP}}$$

Loss increases linearly with the number of terminals on the bus.

Unfortunately, we must also account for excess loss and connector loss L_C (including any loss from diameter and NA mismatches) at each coupler. Since a connector is required at both the input and output ports of the coupler, the number of required connectors is $2N$. These losses also add linearly with the number of terminals, so the real total distribution loss becomes

$$L = (N - 1)\text{Loss}_{\text{THP}} + \text{Loss}_{\text{TAP}} + 2NL_C + \text{Loss}_E$$

As the number of terminals added to a network using tee couplers increases, losses mount quickly. As a result, tee couplers are only useful when a small number of terminals are involved. The difference in losses between an ideal network (having only throughput and tap losses) and a real network (having also excess and connector loss) rapidly becomes greater.

The network in Figure 12-4 is one directional. A transmitter at one end of the bus communicates with a receiver at the other end. Each terminal also contains a receiver. A duplex network can be obtained by adding a second fiber bus. It can also be obtained by using an additional directional coupler at each end and at each terminal. Such additions allow signals to flow in both directions. Figure 12-5 shows an example.

The loss characteristics of a tee network require receivers to have large dynamic ranges. Consider a 10-terminal network with terminal 1 serving as the transmitter. The power received at terminal 2 and at terminal 10 will be quite significantly different. Thus, the dynamic range of the receiver in each terminal must be large. If we assume a coupler with a 10:1 splitting ratio, assume excess and connector losses at each coupler total 2 dB, and ignore all other losses in the system, the difference between receiver power at terminal 1 and at terminal 10 is about 30 dB. The receivers require a dynamic range of 30 dB or better. Other real losses will, of course, increase the requirement.

Notice that failure of a single coupler does not shut down the entire network. It simply divides the network into two smaller networks, one on each side of the failed coupler.

FIGURE 12-5
Tee network
with directional
coupler

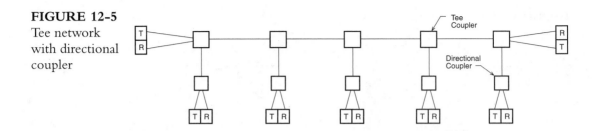

STAR COUPLER

The *star coupler* is an alternative to the tee coupler that makes up for many of the latter's drawbacks. Figure 12-6 shows a transmissive star coupler that has an equal number of input and output ports. For a network of N terminals, the star coupler has $2N$ ports. Light into any input port is equally divided among all the outport ports.

The insertion loss of a star coupler is the ratio of the power appearing at a given output port to that of an input port. Thus the insertion loss varies inversely with the number of terminals:

$$\text{loss}_{\text{IN}} = 10 \log_{10} \frac{1}{N}$$

Loss does not increase linearly with the number of terminals.

If we add excess loss and connector loss, the total distribution loss becomes

$$L = 10 \log_{10} \frac{1}{N} + \text{Loss}_{\text{E}} + 2L_{\text{C}}$$

As a result, star couplers are more useful for connecting a large number of terminals to a network.

Ideally, the light is coupled evenly into all the output ports. Practically, it is not. The actual amount of power into each output port varies somewhat from the ideal determined

FIGURE 12-6
Star coupler

$$\text{Loss}_{\text{IN}} = 10 \log_{10} \frac{1}{N}$$

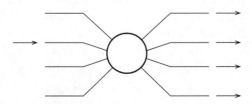

FIGURE 12-7
Loss versus
number of ter-
minals for tee
and star couplers

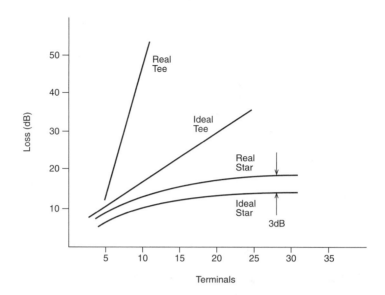

from the insertion loss. *Uniformity* is the term used to specify the variation in output power at an output port. Uniformity is expressed either as a percentage or in decibels. Consider a coupler in which the output power at each port is 50μW. A uniformity of ±0.5 dB means that the actual power will be between 45 and 56 μW. If the uniformity figure increases to ± dB, the output varies between 40 and 63 μW.

Figure 12-7 compares loss with the number of terminals for ideal and real tee and star couplers. The figure assumes an excess loss of 1 dB and a connector loss of 1 dB for each connector, for a total of 3 dB. The figure illustrates the important difference between tee and star couplers. Excess and connector losses occur at each coupler. Since a star network requires only one coupler, these losses occur only once. With a tee network, they increase linearly with the number of couplers in the system. Loss in a star network increases much slower than in a tee network as the number of terminal increases.

One reason for using a tee-coupled network instead of a star-coupled network is that the tee network requires less cable. The centralized location of a star coupler requires significantly more cable to connect widely separated terminals.

Figure 12-8 shows a block diagram of a typical star network using a 4 × 4 star coupler.

REFLECTIVE STAR COUPLERS

A *reflective star coupler* contains N number of ports in which each port can serve as an input or an output. Light injected into any one port will appear at all other ports.

FIGURE 12-8
Star coupler
application

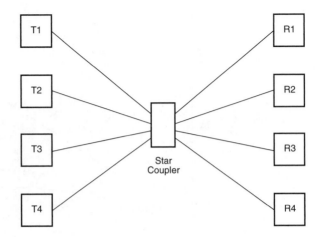

COUPLER MECHANISMS

This section looks at several means of constructing a coupler. Most couplers are "black boxes," in which the internal coupling mechanisms are of little interest to the user. Typical input/output ports to the coupler are either fiber pigtails or connector bushings. The advantage of the pigtail is that it permits greater flexibility of application, since connection to the pigtails can be made with any compatible splice or connector. Couplers using connectorized ports are naturally limited to interconnection by specific connector type. The connectorized couplers are, on the other hand, ready to go; inexperienced users can simply connect or disconnect the fiber to the coupler.

Fused Couplers

A *fused star coupler* is made by wrapping fibers together at a central point and heating the point. The glass will melt into a unified mass so that light from any single fiber passing through the fused point will enter all the fibers on the other side. A transmissive star coupler results when an end of each fiber is on each side of the fused section. A reflective star results when the fibers loop back so that each fiber is twice fused in the fused section. Both types are shown in Figure 12-9.

FIGURE 12-9
Fused star
couplers

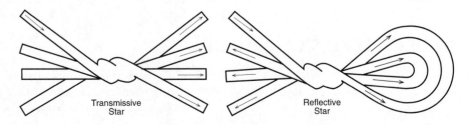

Transmissive
Star

Reflective
Star

Depending on how the fibers are heated and pulled during fusing, the optical energy can be split evenly or unevenly among the fibers. Fused couplers are very small, since the fusing region is only about a tenth of an inch. They also offer very good uniformity.

Planar Waveguide Couplers

A waveguide can be constructed on a silicon substrate using fabrication techniques similar to those used to make semiconductor chips, such as photolithography. The device begins with a substrate, on top of which a waveguide material is deposited. The index of refraction of the waveguide is higher than that of surrounding materials so that the light is guided.

A coupler can be formed in a waveguide in several ways. The simplest is a Y-shaped structure shown in Figure 12-10. Light traveling down the leg of the Y will split onto the two branches. If the angles of the two branches are equal, light will split evenly. Using different angles will result in different coupling ratios between the branches. By cascading Ys, so that a branch (or output) of one coupler becomes the leg (input) of a second coupler, you can achieve any number of outputs.

Another method of constructing a fiber relies on evanescent waves. Evanescent waves can leak out of a waveguide into the surrounding material. If you run a second waveguide

FIGURE 12-10
Planar
waveguide

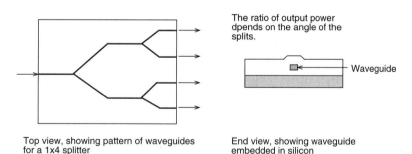

Planar Splitter/Coupler Using Y-Shaped Splits

The ratio of output power dpends on the angle of the splits.

Waveguide

Top view, showing pattern of waveguides for a 1x4 splitter

End view, showing waveguide embedded in silicon

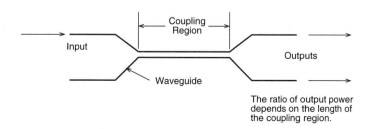

Planar Splitter/Coupler Using Evanescent Waves

Coupling Region

Input

Waveguide

Outputs

The ratio of output power depends on the length of the coupling region.

close and parallel to the first, the energy that leaks out of the first waveguide will transfer to the second. The amount of light coupled from the first waveguide to the second depends on the distance they run parallel (and on the gap separating them). If the parallel run is long enough, all the light will couple from the first to the second waveguide. If the run is even longer, the light will begin to couple from the second waveguide back into the first. Thus, by carefully controlling the distance the two waveguides run closely parallel, you can control the coupling ratio. As with the Y coupler, evanescent-coupled waveguides can be cascaded to achieve different coupling needs.

Compared to an optical fiber, waveguides have high losses. The planar waveguide is very small, however, so that losses are still small because the light travels only a short distance through it.

OPTICAL ISOLATORS

An isolator is a device that permits light to pass through in only one direction. A principle use is preventing reflections in a system. Figure 12-11 shows a typical construction. All surfaces are treated with an antireflective coating to reduce the insertion loss. All surfaces are angled to reduce reflection. Light enters an input fiber and passes through a lens, which collimates the light rays. The light then passes through an isolating element, consisting of a polarizer prism, a Faraday rotator, and an analyzer prism, all in series. Dual-stage isolators include a second isolating element. Following the isolating element, an output lens refocuses the light into the output fiber. Light traveling in the upstream direction is not refocused into the input fiber.

FIGURE 12-11
Optical isolator
(Photo courtesy
of AMP
Incorporated)

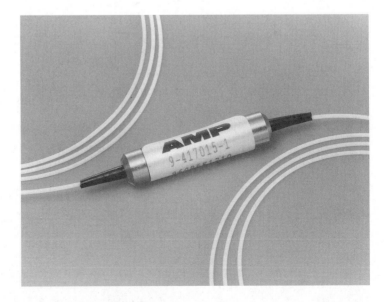

The crystals used in an isolator are birefringent; that is, the materials have a different refractive index for each polarization state. This causes the two polarization states to travel along different paths. The Faraday rotator rotates the optical axis of both polarization states by 45 degrees, affecting the rotation relative to the direction of the magnetic field and not the light's direction of propagation. In the reverse direction, however, the Faraday rotator causes the polarized light to change its polarization state another 45 degrees and become out of phase. The light, therefore, is not refocused into the incoming fiber.

Isolators are often used in laser-based transmitters, at the input and output of optical amplifiers, and other points to prevent reflections.

WAVELENGTH-DIVISION MULTIPLEXER

Multiplexing is a method of sending several signals over a line simultaneously. In Chapter 2, we saw how telephone companies use time-division multiplexing to send hundreds of telephone calls over a single line. Each voice is first transformed in digital data by pulse-coded modulation (PCM). Each PCM-encoded signal is then allotted specific time slots within the transmission. Actual modulation and demodulation are accomplished electrically, before being presented to the fiber-optic transmitter and after being received from the fiber-optic receiver.

Wavelength-division multiplexing (WDM) uses different wavelengths to multiplex two or more signals. Transmitters operating at different wavelengths can each inject their optical signals into an optical fiber. At the other end of the link, the signals can again be discriminated and separated by wavelength. A WDM coupler serves to combine separate wavelengths onto a single fiber or to split combined wavelengths back into their component signals.

Figure 12-12 shows an example of WDM. Two transmitters, one operating at 850 nm and one at 1300 nm, present signals to a WDM coupler, which couples both signals onto a fiber. At the other end, a second WDM coupler separates the received light back into its 850- and 1300-nm components and presents it to two receivers.

Two important considerations in a WDM device are crosstalk and channel separation. Both are of concern mainly in the receiving or demultiplexing end of the system. *Crosstalk* or directivity refers to how well the demultiplexed channels are separated. Each channel should appear only at its intended port and not at any other output port. The

FIGURE 12-12
Wavelength-division multiplexing

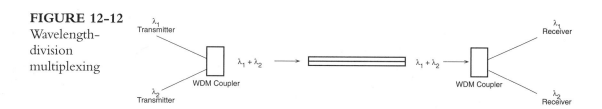

crosstalk specification expresses how well a coupler maintains this port-to-port separation. Crosstalk, for example, measures how much of the 850-nm wavelength appears at the 1300-nm port. A crosstalk of 20 dB means that 1% of the signal appears at the unintended port.

Channel separation describes how well a coupler can distinguish wavelengths. In most couplers, the wavelengths must be widely separated, such as 850 nm and 1300 nm. Such a device will not distinguish between 1290-nm and 1310-nm signals.

Most WDM devices today operate with only a few separate optical channels. Practical, affordable couplers do not discriminate between closely separated channels. A three-port coupler, for example, might multiplex 850-nm and 1300-nm wavelengths, whereas a four-port coupler multiplexes 755-, 850-, and 1300-nm wavelengths.

The limited number of channels for WDM form the spectral requirements of a multi-channel WDM device. LED sources have wide spectral widths, which require each channel to be separated widely from the others. Constructing a WDM device with many channels requires a narrow band source. Lasers, which have sufficiently narrow spectral widths allow many channels in the same optical window.

Popular two-wavelength multiplexing schemes include:

850/1300 nm
1300/1550 nm
1480/1550 nm
985/1550 nm

WDM allows the potential information-carrying capacity of an optical fiber to be increased significantly. The bandwidth-length product used to specify the information-carrying capacity of a fiber applies only to a single channel—in other words, to a signal imposed on a single optical carrier. A fiber with a 500-MHz-km bandwidth used with a four-port WDM coupler can carry a 500-MHz signal on the 755-nm channel, a 500-MHz signal on the 850-nm channel, and a 500-MHz signal on the 1300-nm channel. Its effective information-carrying capacity has been increased to 1500 MHz.

DENSE WAVELENGTH-DIVISION MULTIPLEXERS

A more recent innovation in optical multiplexing is the dense WDM, which multiplexes and demultiplexes wavelengths in the same window. Widely used in telecommunications applications to increase capacity, the dense WDMs can distinguish between wavelengths separated by less than 1 nm. Because dense WDMs are so important to telecommunications applications, we will discuss them further in Chapter 16. For now, we will describe basically how they work.

FIGURE 12-13
Using optical filters for optical demultiplexing

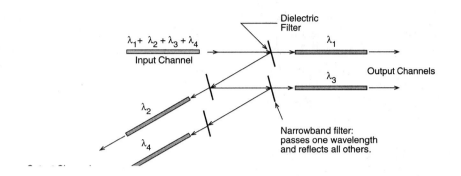

Figure 12-13 shows one example of how a DWDM can be constructed for demultiplexing. At the heart are a series of narrow bandpass filters, each for a different wavelength. Let's say that the incoming light carries four channels at 1550, 1552, 1554, and 1556 nm. The light first passes through a lens that collimates it and presents it to the first filter. This filter passes one wavelength (say, 1550 nm) to another lens that focuses the passed light into an output fiber. The first filter reflects the rest of the light to a second filter. This filter passes a second channel (say, 1552) to a second output fiber and reflects all other wavelengths. This passing-one-wavelength-and-reflecting-all-others continues for as many channels as there are in the incoming signal. The light cascades from filter to filter, with each filter removing a single channel.

Many DWDM designs are bidirectional. The example shown in Figure 12-13 shows a demultiplexing application. If you simply reverse the arrows, you can see how the device serves to multiplex many wavelengths onto a single fiber. The bandpass filter passes the wavelength of interest in either direction. Other filters reflect the light so that the light is continuously combined at each stage of the device.

It should be obvious that the filter must be very precise, passing only the selected wavelength and no others. Light "leaking" through a filter is a form of crosstalk: It both adds unwanted light to a channel and robs energy from the channel it belongs to. In other words, you don't want 1554-nm signals output in the 1550-nm channel. And you don't want the power in the 1554-nm channel reduced through crosstalk. Fiber Bragg gratings and other approaches can also be used to construct DWDM systems.

Another area of interest is maintaining low insertion loss. With all the reflections of light, a poorly designed DWDM can rob the optical channels of significant amounts of power. However, DWDMs are becoming increasingly efficient so that losses are controlled and acceptably low, only a few decibels for a multichannel device. As we will discuss shortly, optical amplifiers also are available to amplify the optical signal.

To allow such closely placed channels requires laser sources with very narrow spectral widths that can be tuned to a precise wavelength. Distributed-feedback lasers have a spectral width of around 0.2 nm or less, and so they fit the need. In addition, the DWDM must use optical gratings or other wavelength-sensitive devices to ensure that only the proper wavelength is demultiplexed onto each output fiber. Since optical filters are a critical part of DWDMs, the next section discusses approaches to filters.

OPTICAL FILTERS

You're already familiar with an optical filter. Sunglasses are filters in a broad sense. If you are a photographer, you probably use filters. A red lens, for example, passes reddish light and either absorbs or reflects others. The range of wavelength passed is typically broad. Optical filters in fiber optics work similarly, but with greater precision. A filter can act as either a bandpass filter or a band-reject filter. The bandpass filter passes only a selected range of light and rejects all others. A band-reject filter does the opposite: It rejects a selected range and passes others.

Optical filters in fiber systems can have various bandwidths: as narrow or wide as the application requires. A wideband filter is typically used to distinguish between widely separated wavelengths, such as between 1310 nm and 1550 nm. The bandpass may be as wide as 40 nm, meaning that a filter designed to operate at 1550 nm will pass most of the light between 1530 and 1570 nm.

Other filters are narrowband filters that must be selective to less than 1 nm in some cases. The newest telecommunications systems require narrowband filters that can distinguish between wavelengths that are only 0.8 nm apart: It must be able to pass a 1551-nm wavelength while rejecting a 1551.8-nm signal. The filter can operate by attenuating signals or reflecting them. Reflection is preferred in some applications because the reflected signal will still contain signals that will be selected by subsequent filters.

Dielectric Filters

Depositing thin layers on a glass substrate allows an optical filter to be constructed that transmits only one wavelength and reflects all others. Dielectric filters can be useful for multiplexing and demultiplexing a low to medium number of wavelengths. Each filter also presents low insertion loss, but because a multiwavelength device is formed by bouncing the light from one filter to the next, optical loss becomes varied with the number of channels. Upgrades to increase the number of wavelengths are easily accomplished, but the channel count is limited to about 16 wavelengths.

Figure 12-13, above, showed how a DWDM can be constructed using dielectric filters. The figure shows demultiplexing applications. If you reverse the direction of the arrow, you can see how it operates as a multiplexer.

Optical Bragg Grating

During the mid-90s, fiber Bragg gratings (FBGs) became a significant filtering technology. While the technology has been around about ten years, its commercial application has only recently occurred. A Bragg grating can be built directly into a fiber to act as a wavelength-selective filter. An FBG is made by subjecting a short length of fiber to strong ultraviolet light. A mask is used to form a pattern through which the UV light passes (Figure 12-14). The strong light, typically in the 240- to 260-nm region, creates interference patterns that permanently alter the refractive index of the fiber core at precisely defined intervals. The mask is precisely made in relation to the wavelengths of light it will deal with. The patterns of altered refractive indexes are periodic; that is, they repeat in regular patterns. Different periodic patterns can be combined to work on a variety of wavelengths.

When a stream of light reaches the FBG, most wavelengths pass through with little loss and no disruption to the signal. However, certain wavelengths will be strongly reflected. The wavelengths reflected depend on the period of the grating: The grating will reflect wavelengths that are twice the grating period. In other words, to reflect a 1550-nm wavelength, the grating should have a period of 775 nm. The operation of the FBG depends on three things: the strength of the changes in refractive index, the length of the grating (how many periods), and *the precision with which the gratings are formed.*

Figure 12-15 shows a curve for a typical FBG used as a narrowband filter. The wavelength of 1552 nm is strongly reflected, while other wavelengths pass through with next to no insertion loss. The transmission loss of nearly −50 dB shows that most of the light at this wavelength is reflected. The 1552-nm light passing through the grating is −50 dB less than the 1552-nm optical power into the grating.

FBGs can be constructed with many different properties to meet application needs. Some select the narrowest of wavelength bands and offer extremely strong reflections. FBGs are available with bandwidths of under 0.2 nm and the ability to reflect over 99.9%

FIGURE 12-14

Constructing a fiber Bragg grating filter (Courtesy of Ibsen Microstructures)

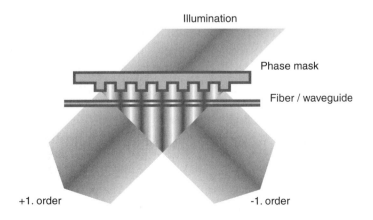

FIGURE 12-15
FBG filter
response

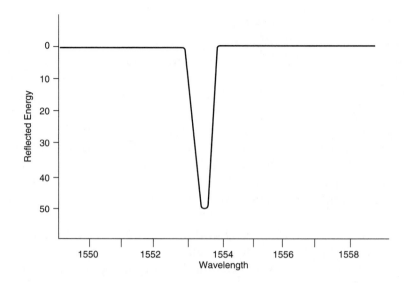

of the light. In this case, bandwidth refers to the width of the light reflected. For a 1550-nm signal, it will strongly reflect light from 1549.9 to 1550.1 nm. Others work over a wider band and have weaker reflections or even couple light from the core into the cladding. For example, you may want to flatten the power in one wavelength compared to another. In this case, you might want the FBG to reflect only 10% of the energy at a specific wavelength and pass the rest. The precision with which an FBG can be built to select only very narrow wavelengths makes it popular in WDM and other applications that require very narrow wavelength selection. FBGs have been used for the following:

- Wavelength multiplexers and demultiplexers in dense wavelength-division multiplexing. FBGs can have a resolution of well under 1 nm. Figure 12-16 shows two approaches to using an FBG for DWDM applications. One uses an optical circulator, which is a device that sequentially presents the light to each optical port. At port 2, all the wavelengths pass except the selected one, which is reflected. This wavelength then appears as output on port 3. The circulator-based approach is simple, but the cost of circulators tends to be high at this point. The second approach is based on what is known as a Mach–Zehnder interferometer, which uses a coupler-like structure with waveguides. It is more complex, but avoids the circulator.

- Dispersion compensation: a chirped FBG can be used to recompress pulses back to their original shape. Essentially, the FBG negates chromatic dispersion. It does so by varying the period along the length of the grating. While a WDM grating reflects only one wavelength, an FBG for dispersion compensa-

FIGURE 12-16
Two approaches
to using FBGs in
DWDM applica-
tions

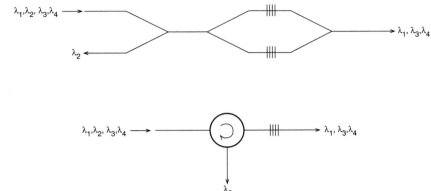

tion reflects different wavelengths along the length of the grating. This introduces differing amounts of delay for different wavelengths. By carefully gaug-
ing the delays needed in terms of the dispersion, you can build a filter that
"tightens" the pulse.

- Wavelength locking of pump lasers in optical-fiber amplifiers. The FBG
  ensures the narrow spectral width of the output.
- Gain-flattening filter in optical-fiber amplifiers. This ensures that all wavelengths are amplified an equal amount to the same amplitude. A long-period
  FBG couples light from higher-gain wavelengths into the cladding. All wavelengths show the same gain at the output.

FBGs offer excellent wavelength selectivity and filter shape; that is, they cleanly distinguish between different wavelengths. In addition, the technology is very flexible and
allows great latitude in choosing the type of filter created for different needs. Drawbacks
include temperature dependence, which requires the device to be maintained at a consistent temperature through heating or cooling. In addition, filtering a number of wavelengths can create complex structures.

Arrayed Waveguide Grating

An arrayed waveguide grating is similar to the silicon-based coupler we discussed earlier. Waveguides are formed on a planar substrate. The AWG works on the interference
effect between different waveguides. A series of curved waveguides are formed, with each
waveguide having a progressively longer length. At each end of the array is a slab that
works as a radiative star coupler. As light travels through the waveguide, it interferes with
the light in other waveguides. In effect, it is phase-shifted in relation to other light. The
output slab from the array acts like a diffraction grating, separating the light by wave-

lengths and directing each wavelength to a different output waveguide. The center wavelength, the channel spacing (difference of wavelength in each waveguide), and the number of channels is determined by the length of one of the waveguides, the change in length of subsequent waveguides, and the spacing between them. Figure 12-17 shows the construction of an AWG.

An advantage of the AWG is that cost is not as dependent on the number of channels. It also offers uniformity of power across a large number of channels. The operation is, however, sensitive to changes in temperature and the AWG must be heated or cooled to maintain a constant temperature.

Figure 12-18 is a table comparing the various approaches. The filter shape refers to how sharply and cleanly the filter can distinguish between wavelengths—it measures the quality of the filter in separating closely spaced wavelengths. Temperature dependence shows to what degree the filtering technique is affected by temperature changes. The cost of adding channels refers to the incremental costs of adding additional wavelengths. The final column, *loss per number of wavelengths,* shows the degree to which loss increases with an increasing number of wavelengths. Bear in mind, too, that these technologies are fairly new in fiber optics and improvement in technology may change their relative merits. For example, the filter shape of the AWG can be improved as the processing and fabrication technology improves.

OPTICAL FIBER AMPLIFIERS

Since a signal becomes attenuated as it travels along a fiber, it must be periodically strengthened. The traditional method has been to convert the optical signal into an electri-

FIGURE 12-17
Arrayed waveguide grating
(Courtesy of Photonics Integration Research, Inc.)

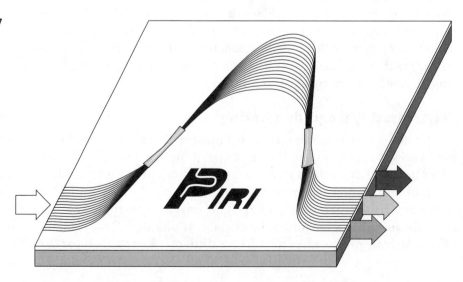

FIGURE 12-18
Comparison of
filter techniques
for DWDM

FILTER TECHNIQUE	FILTER SHAPE	TEMPERATURE DEPENDENCE	COST OF ADDING CHANNELS	LOSS WITH INCREASING CHANNELS
Dielectric Filter	Very good	Low	High	High
Fiber Bragg Grating	Excellent	High	High	Medium
Arrayed Waveguide Grating	OK	High	Low	Low

cal signal, amplify it, and reconvert it back into an optical signal. This process is commonly called regeneration. Optical amplifiers allow optical signals to be amplified directly, without conversion. The most common way to do this is with an optical fiber amplifier (OFA). The most popular type of OFA is called an erbium-doped fiber amplifier (EDFA).

A short length of fiber that has a small amount of the element erbium added during manufacture can act as an amplifier. See Figure 12-19 for how it works. Consider a fiber operating at 1550 nm. A short length of erbium-doped fiber is spliced into the fiber. A three-port WDM is also added. One input port of the WDM carries the 1550-nm signal. The other is attached to a pump laser operating at either 980 nm or 1480 nm. The pumping light energizes the erbium. When the energized erbium loses its extra energy, it transfers it to the 1550-nm signal, amplifying it up to 30 dB.

OFAs can be applied in three ways:

1. At the transmitter to increase output power.
2. In midspan to replace regenerative repeaters.
3. At the receiver to increase sensitivity by acting as a preamp.

FIGURE 12-19
Optical fiber
amplifier

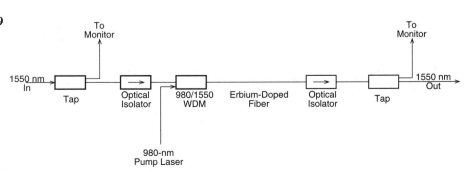

In the local loop, OFAs promote efficient branching, allowing the signal to be split many times while maintaining adequate power levels.

OFAs can use a single-pumped and dual-pumped configuration. The single-pumped OFA uses a single pump laser on the upstream side of the doped fiber. The dual-pumped version uses a pump laser on both ends. Typical gains for the amplifiers are +17 dB for a single-pumped amplifier and +35 dB for a dual-pumped version. A typical 1550-nm transmitter used in long-haul applications has an output power of −5 dBm for digital and +5 to +7 dBm for analog. Amplified signals can be as high as +30 dBm for digital and +42 dBm for analog, allowing optical energy to be tapped off many times in branching applications and making fiber-to-the-curb or -home more practical.

The choice between a 980-nm and 1480-nm pump laser involves tradeoffs in performance versus reliability. Lasers operating at 1480 nm have traditionally been more reliable because they are similar in structure to 1550-nm lasers. The 980-nm laser, however, has features that make it more attractive. First, it has greater quantum efficiency and requires less drive current to achieve a particular output level. For example, to achieve 100 mW of output power, the 980-nm laser might require 225 mA while the 1480-nm device requires 400 mA. Second, the 980-nm laser contributes less noise to the system. Third, 980-nm are available with fiber Bragg gratings to stabilize the optical output.

SUMMARY

- A coupler is used to distribute light.
- The two principal types of couplers are tee couplers and star couplers.
- A network is a multipoint fiber-optic system.
- A tee coupler is most useful in a network containing few terminals.
- A star coupler is most useful in a network containing many terminals.
- Wavelength-division multiplexing is a method of multiplexing two or more optical channels separated by wavelength.
- A fiber-optic switch permits light from one fiber to be switched between two or more fibers.

Review Questions

1. Name the two main types of couplers.
2. Sketch the operation of a three-port directional coupler.
3. Explain why a tee coupler is more useful when used in a network with few terminals. Give two advantages of a tee coupler over a star coupler.
4. Calculate the loss for a 4 × 4 star coupler connected to four terminals, and for an 8 × 8 star coupler connected to eight terminals. Ignore excess and connector losses.
5. What is the difference between transmissive and reflective star couplers?
6. Assume a fiber has a 600-MHz-km bandwidth and, through WDM, transmits optical channels at 850 nm and 1300 nm. For a 1-km link, what is the highest bandwidth of each channel? Why?
7. What is the total information-carrying capacity of the fiber described in Question 6?

PART THREE

FIBER-OPTIC
SYSTEMS

chapter thirteen

THE FIBER-OPTIC LINK

We have now looked at the main components of a fiber-optic link, at cables, sources and transmitters, detectors and receivers, and connectors and couplers. This chapter looks at how these components are brought together into a link and at how a link is planned. It details the ingredients of a power budget and a rise-time budget, which are the two fundamental requirements that ensure that a link meets its intended application requirements. We will learn how to construct and evaluate such budgets for a simple fiber-optic system.

PRELIMINARY CONSIDERATIONS

The first step in any application is to have an application suitable to fiber optics in the first place. Wiring your doorbell with a fiber-optic link might be an interesting exercise, but it is hardly practical. Similarly, linking a personal computer to a nearby printer is most easily and inexpensively done with a standard electrical cable. If, however, the printer is a great distance away or there are strong noise sources nearby that interfere with transmission, fiber optics becomes of interest.

The decision to use fiber optics involves comparing its advantages, disadvantages, and costs against competing copper solutions. We will, however, assume fiber optics has been chosen as the best medium.

The next step is to decide whether to buy a complete fiber-optic system, to use packaged transmitters and receivers and build your own system from there, or to design and build your own transmitters and receivers as well. Because of our interests in this chapter, we select the middle choice. The third option sidetracks us from the main purpose of this chapter, which is budgeting the optical portion of the link. The first option, which will be installed by a vendor or contractor, leaves us with nothing to discuss.

SYSTEM SPECIFICATIONS

In planning a fiber-optic system, we must define our application requirements so that we can specify our needs. The main question involves the data rates and distances involved: How far? How fast? These questions give us the basic application constraints. Beyond that, we must specify the BER required.

Now that we have the main requirements—distance and data rate—we can begin to evaluate the other factors involved:

- Type of fiber
- Operating wavelength
- Transmitter power
- Source type: LED or laser
- Receiver sensitivity
- Detector type: pin diode, APD, IDP
- Modulation code
- BER
- Interface compatibility
- Number of connectors
- Number of splices
- Environmental concerns
- Mechanical concerns
- Other special concerns

The environmental and mechanical concerns involve such issues as temperature and humidity ranges, indoor/outdoor application, flammability requirements, and so forth. These factors will especially affect the choice of fiber-optic cable.

We can see that many of these questions are related and cannot be as easily separated as we have done here. The receiver sensitivity is influenced by the choice of detector. The receiver sensitivity sets the minimum optical power required by the receiver. The power arriving at the receiver, however, depends on the transmitter power and the fiber attenuation. We can specify a very sensitive receiver that will allow us to use a transmitter of lower power. Or we can use a less sensitive receiver, in which case our transmitter must be more powerful.

Consider a given transmitter, fiber-optic cable, and receiver. Now assume the power at the receiver is insufficient to meet the BER requirement of the application. We actually have five choices to remedy the situation:

1. A transmitter with a higher output power
2. A fiber with lower attenuation
3. A shorter transmission distance, to lower losses along the fiber length
4. A receiver with a lower sensitivity level
5. A modification of the system requirements, so that a lower BER is acceptable (a lower BER implies a lower minimum power at the receiver)

The point is that planning a fiber-optic link is not a cut-and-dried, step-by-step procedure. However, there are logical and rational ways to proceed. One approach is the link power budget.

POWER BUDGET

The power budget is the difference between *minimum* transmitter output and *minimum* receiver sensitivity. It defines the worst-case link loss by accounting for the least power out of the transmitter and the minimum received power. Let's look at an actual example—Fast Ethernet for multimode cable. As you will see in Chapter 15, Fast Ethernet is a type of network operating at a 100-Mbps data rate and a 125-Mbps transmission rate using 8B/10B encoding. The transmitter output specified by the standard is from −14 dBm maximum to −19 dBm minimum. The receiver sensitivity is −33.5 dBm, which is the minimum power to the receiver required to maintain an acceptable BER. The difference between the minimum output power and the minimum sensitivity is 14.5 dB: −33.5 dBm − (−19 dBm) = 14.5. The power budget is 14.5 dB. The losses in the transmission path between transmitter and receiver cannot exceed 14.5 dB if the system is to achieve an acceptable BER.

In an actual Fast Ethernet system, the output power is usually greater than −19 dBm. A typical budget is around 20 dB, or 5.5 dB greater than the power budget.

Once you know the power budget, you can calculate losses throughout the system to ensure the budget is not exceeded. Essentially, this calculation involves three things:

- Interconnection losses from connectors and splices.
- Fiber attenuation. You can generally assume loss is linear with length. Thus, if the cable has a loss of 3.5 dB/km, a 2-km link has a loss of 7 dB.
- A power margin to account for aging of components (remember the output of an LED declines over time), repairs, and other changes that may occur over the life of the system. Power margins range from 3 to 6 dB. The 3-dB figure suffices for most applications.

To return to our 14.5-dB power margin for Fast Ethernet, if we subtract the 3-dB loss for our margin, we have 11.5 dB left for losses. For each connector, assume a loss of 0.75 dB (since this is the maximum loss allowed by premises cabling and network specifications). Most connectors have a loss of 0.2 to 0.3 dB, so again we are building in a margin. Use 0.3 -dB for each splice's insertion loss.

Figure 13-1 shows a typical configuration for Fast Ethernet. An Ethernet hub (containing a transmitter and receiver) connects through a patch panel to the building cabling. It runs 300 meters to a wall outlet in an office or work area. A splice at an intermediate point divides the cable running between floors from the cable running along one floor of the building. The losses become:

2 patch panel interconnections (@ 0.75 dB):	1.5 dB
1 splice	0.3 dB
300 meters of fiber (3.5 dB/km @ 850 nm)	1.05 dB
Total	2.85 dB

Of the 11.5 dB budget, we still have 8.65 dB remaining. We haven't included the two patch cables (one from the hub to the patch panel and one from the work outlet to the computer). These are less than 10 meters each, so their loss is negligible and can be safely ignored since we have a lot of margin left.

Also shown in Figure 13-1 is a graphical depiction of the system loss. Power is shown on the vertical axis; distance is shown on the horizontal axis. Power levels are shown at different points in the system.

Typically, you do not have to calculate connection losses in attaching the fiber to the transmitter or receiver. The transmitter is specified as launching a given amount of power into a given type of fiber. The receiver sensitivity is similarly specified with respect to a

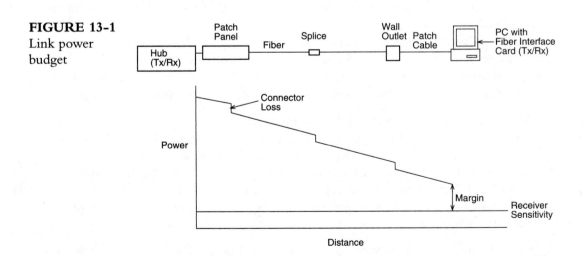

FIGURE 13-1
Link power budget

specific type of fiber. For multimode fiber, 62.5/125-μm fiber is typically specified, although 50/125-μm fiber is sometimes used. If you use a different type of fiber, the amount of power coupled into it may change. The received power may also change.

Basic power budgets, you can see, are fairly straightforward. There are three considerations you must keep in mind.

First, is the relation between bandwidth and loss. A link can be bandwidth-limited or loss-limited. A loss-limited link is one in which losses determine the link limits. For example, for the Fast Ethernet, we can extend the distance to nearly 3 km before we reach the limits set by the power budget. In a bandwidth-limited link, the length is limited by the bandwidth of the fiber. In a multimode system with 62.5/125-μm fiber, the distance is limited to 1.6 km. This is because the fiber has a bandwidth of 160-MHz-km. So, running at 850 nm over 62.5/125 fiber does not allow a 3 km length because the link is bandwidth-limited. Therefore, the link at 850 nm is bandwidth limited.

At 1300 nm, the fiber's bandwidth increases to 500 MHz-km and its loss decreases to 2 dB/km. If you do a quick calculation, you find the link limits become:

Loss: 4.85 km
Bandwidth: 5 km

The link will be limited, in this case, by loss. The maximum distance is about 4.85 km.

The second aspect is that most simple budgets assume you are using the same type of fiber throughout the system. Most current standards recommend 62.5/125-μm fiber, although 50/125-μm fiber is becoming popular because of its lower loss, higher bandwidth, and compatibility with newer sources such as VCSELs. When different types of fibers are used, additional losses from diameter mismatches and NA mismatches must be factored into the budget. When light passes from a 50-μm core to a 62.5-μm core, no losses from mismatches occur. When the opposite occurs—the larger 62.5-μm core transmitting into a smaller 50-μm core—significant losses occur. As you saw in Chapter 11, the mismatches cause an NA-mismatch loss of 2.8 dB and a core-diameter mismatch of 2.9 dB. The total loss is 5.7 dB. Obviously, you want to avoid using different fiber sizes in the link.

Earlier in this chapter, we said that you can usually ignore patch cable because of their short size and negligible loss. However, keep in mind that patch cords can be a significant consideration if they cause a mismatch. If you have a 62.5/125-μm fiber in the building cabling and use 50-μm fiber for patch cables, you will have significant loss at the receiving end of the system. Similarly, with a 50/125-μm cabling system and 62.5/125-μm patch cables, the mismatch loss will be at the transmitting end.

The third thing to remember is that most standard applications specify everything you need to know about the link, including bandwidths, distances, fiber types, and transmitter and receiver characteristics. Again, for Fast Ethernet, the network hub and interface card in the PC should meet all transmitter and receiver requirements. The type of cable is specified and the recommended distances are specified. All you have to do is ensure that the recommendations are met. It's rather straightforward, however, at times, you might want to install a nonstandard system or even stretch a standard.

RISE-TIME BUDGET

The power budget analysis ensures that sufficient power is available throughout the link to meet application demands. Power is one part of the link requirement. The other part is bandwidth or rise time. All components in the link must operate fast enough to meet the bandwidth requirements of the application. The rise-time budget allows such analysis to be performed.

As we saw in the chapters on sources and detectors, active devices have finite response times in response to inputs. The sources and detectors do not turn on and off instantaneously. Rise and fall times determine response time and the resulting bandwidth or operating speed of the devices.

Similarly, dispersion limits the bandwidth in the fiber. In a single-mode fiber, only material dispersion from the source spectral width severely limits bandwidth. In a multimode fiber, the delay caused by light traveling in different modes also limits bandwidth or cable rise time.

When the bandwidth of a component is specified, its rise time can be approximated from

$$t_r = \frac{0.35}{\text{BW}}$$

This equation accounts for modal dispersion in a fiber. The rise time must be scaled to the fiber length of the application. If the cable is specified at 600 MHz/km, and the application is 2 km, the bandwidth is 300 MHz and the rise time is 1.6 ns.

In addition, the rise-time budget must include the rise times of the transmitter and receiver, which are typically specified for packaged devices. Transmitter and receiver rise times are used instead of source and detector rise times, since the transmitter and receiver circuits will limit the maximum speed at which opto-electronic devices can operate. For the receiver, rise time/bandwidth may be limited by either the rise time of the components or by the bandwidth of the *RC* time constant.

Connectors, splices, and couplers usually do not affect system speed, and they do not have to be accounted for in the rise-time budget.

When all individual rise times have been found, the system rise time is calculated from the following:

$$t_{\text{rsys}} = 1.1 \sqrt{t_{r1}^2 + t_{r2}^2 + \ldots + t_{rn}^2}$$

The 1.1 allows for a 10% degradation factor in the system rise time. In a typical application, the rise-time budget can also be used to set the rise time of any individual components, such as the transmitter or cable, by rearranging the equation to solve for the unknown rise time.

When rearranging the equation to solve for an individual rise time, the 1.1 factor is not used. Since, after all, the factor allows for degradation within the entire system, it should be applied to individual components.

Consider as an example a 20-MHz application operating over 2 km. The fiber used has a 400-MHz-km bandwidth. The receiver rise time is 10 ns. What is the rise time required of the transmitter?

The required system rise time is 17.5 ns. Fiber rise time is 1.75 ns. Solving for the transmitter rise time gives

$$t_{rtrans} = \sqrt{17.5^2 - 10^2 - 1.75^2}$$

$$= 14.25 \text{ ns}$$

The transmitter must have a rise time of about 14 ns. If we select a transmitter with a rise time of 10 ns, we will meet the rise time or bandwidth requirements of the application. With a 10-ns transmitter, the system rise time becomes

$$t_{rsys} = 1.1\sqrt{10^2 + 1.75^2 + 10^2}$$

$$= 15.7 \text{ ns}$$

which is within the required 17.5 ns.

FIBER NETWORK DESIGN SOFTWARE

Fiber-optic networks can also be designed and analyzed with computer software. Bear in mind that the examples in this chapter are rather simple. A network can involve hundreds of users in multiple buildings on a campus. In essence, the network may contain hundreds of links, various interconnections, and different types of fiber-optic cable (single-mode between buildings and multimode within a building is typical). In addition to ensuring each link will operate properly, you need a way to keep track of everything once the network is built.

With a software design program, such as FiberGrafix Network Design Software for Windows, it is possible to graphically depict the entire network and generate a complete list of cable and apparatus products required to build it. Furthermore, using mathematical models developed by Bell Laboratories and incorporated into the software, it is possible to

FIGURE 13-2
Software can aid system design (Courtesy of Lucent Technologies)

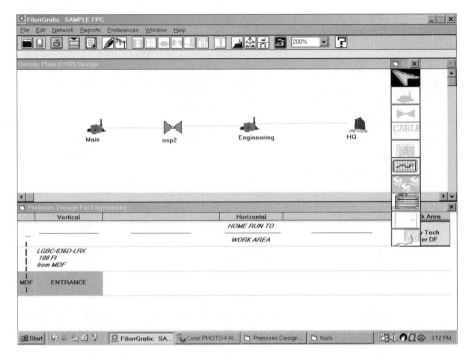

verify that the network will operate within the transmission characteristics of the selected opto-electronics. Figure 13-2 shows a screen image of a campus network designed using the FiberGrafix software. The upper window shows the outside (campus) portion of the network and the lower window shows 54 work areas within the "Medical Center" building connected with fiber-to-the-desktop in a traditional "hierarchical star" architecture.

To design a long-haul, campus, or premises fiber-optic network such as shown in this example, the user would follow these logical steps:

1. Define your fiber system type (multimode or single-mode). The decision between single-mode and multimode cables is fundamental, since it will be key to other decisions concerning connectors, connecting hardware, and so forth.
2. Place buildings, premises locations, and splice points using a mouse and the software's "drawing pad."
3. Connect buildings, premises locations, and splice points with cable (specifying environment, length, and number of fibers for each cable).
4. Specify what to do with each fiber in each cable (splice, interconnect, cross-connect, and patch panels, closures, outlets, etc.).

5. Select opto-electronics required for your application—for example, Ethernet, Fast Ethernet, Token Ring, ATM, "generic," etc.
6. Perform the Optical Path Analysis to verify operation within capabilities of the opto-electronics.
7. Generate reports containing lists of cable and apparatus products required to build the network.
8. (Optional) Perform "what if" analysis to determine effect of changing something in the design, e.g., splice location, cable type, etc.

The various "tools" used to design the fiber-optic network are represented by various icons, as shown in Figure 13-2. Since the program shown in Figure 13-2 is offered by Lucent Technologies, it has a database of Lucent fiber-optic cable, connectors, and other hardware products and even allows new products to be added periodically through software updates available on the Internet. The software also contains many opto-electronic products, such as fiber network hubs and switches, from a number of vendors. The large number of fiber cable and hardware products, as well as the wide variety of opto-electronics, allows users to design many different types of fiber-optic networks. (The approach, incidentally, also helps insure compatible products since it is generally easier to pick a product from the database than specify a new one where the user is responsible for the "configuration rules.")

As will be discussed in Chapter 15, there are many approaches to building a fiber-optic network. A good design program will make it easy to build networks using a wide variety of architectures, such as hierarchical star, centralized fiber-optic cabling, etc., to simplify your job.

In addition to fiber network design software, cable management software packages are available to help you document your system after it is designed and built. In any system but the most simple, it is very important to keep track of what's connected to what. Documentation is essential—without documentation, you could be in trouble when the only person who understands how everything is connected finds another job. These cable management packages provide a convenient way to document these systems once they are installed.

SUMMARY

- A power budget ensures that losses are low enough in a link to deliver the required power to the receiver.
- A rise-time budget ensures that all components meet the bandwidth/rise-time requirements of the link.

Review Questions

1. Perform a link budget analysis to determine the overall fiber length possible. Assume the following:

   Transmitter output: −14 dBm
   Receiver sensitivity: −33 dBm
   Fiber: 50/125, with 600-MHz-km bandwidth and 4.0-dB/km loss
   Connectors: one interconnection every 2 km; 1-dB loss per connection
   Margin: 6 dB

   Assume the transmitter power and receiver sensitivity accounts for connectors. In other words, you do not have to calculate losses at the transmitter or receiver.

2. Sketch a power budget graph for Question 1.

3. Repeat the analysis from Question 1 for a 100/140 fiber with a bandwidth of 400-MHz-km and loss of 6 dB/km.

4. Sketch a power budget graph for Question 3.

5. Determine the rise-time budget for the link in Question 1. Use the length of fiber determined in the budget. Assume transmitter and receiver rise times of 3 ns.

6. How does network design software ease the task of performing power and rise-time budgets?

chapter fourteen

FIBER-OPTIC CABLE INSTALLATION AND HARDWARE

This chapter looks at some of the factors involved in installing a fiber-optic cable system. These factors are of interest because they not only demonstrate the practical importance of the mechanical properties of a cable, but they also demonstrate the practical aspects of dealing with fiber optics. We will also describe some types of common fiber-optic hardware, such as closure/organizers, rack boxes, and distribution panels. This hardware is an important part of more complex fiber-optic systems.

BEND RADIUS AND TENSILE RATING

Because of their light weight and extreme flexibility, fiber-optic cables are often more easily installed than their copper counterparts. They can be more easily handled and pulled over greater distances.

Minimum bend radius and *maximum tensile rating* are the critical specifications for any fiber-optic cable installation, both during the installing process and during the installed life of the cable. Careful planning of the installation layout will ensure that the specifications are not exceeded. In addition, the installation process itself must be carefully planned. This chapter discusses the factors involved in the planning.

The minimum bend radius and maximum tensile loading allowed on a cable differ during and after installation. An increasing tensile load causes a reversible attenuation increase, an irreversible attenuation increase, and, finally, cracking of the fiber. The tensile loading allowed during installation is higher than that allowed after installation. Care must be taken in either case not to exceed the specified limits.

The minimum bend radius allowed during installation is larger than the bend radius allowed after installation. One reason is that the allowed bend radius increases with tensile loading. Since the fiber is under load during installation, the minimum bend radius must be larger. The allowed bend radius after installation depends on the tensile load.

Figure 14-1 shows the cross section of a simplex fiber and a duplex fiber, which we will use as examples in this chapter, and gives the specifications for their minimum bend radius and maximum tensile rating.

FIGURE 14-1
Bend radius and tensile strength loading for cable examples (Courtesy of Canoga Data Systems)

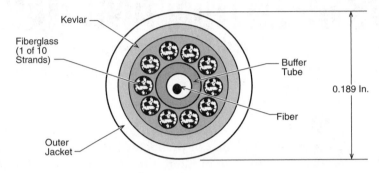

Simplex Cable Cross Section

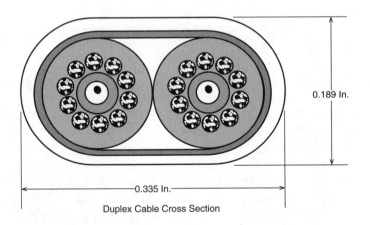

Duplex Cable Cross Section

Tensile Loading (Max)	Cable
During After	90 lb (400 N) 11.25 lb (50 N)
Bend Radius (Min)	**Cable**
During After (No Load) After (Full Load)	5.9 in (150 mm) 1.2 in (30 mm) 4.0 in (130 mm)

As discussed in Chapter 7, outdoor cables are commonly multifiber cables with constructions more complex than indoor cables. Indoor cables are typically simplex or duplex.

DIRECT BURIAL INSTALLATION

Cables can be buried directly in the ground by either plowing or trenching methods. The plowing method uses a cable-laying plow, which opens the ground, lays the cable, and covers the cable in a single operation. In the trench method, a trench is dug with a machine such as a backhoe, the cable is laid, and the trench is filled. The trench method is more suited to short-distance installations.

Buried cables must be protected against frost, water seepage, attack by burrowing and gnawing animals, and mechanical stresses that could result from earth movements. Armored cables specially designed for burial are available. Cables should be buried at least 30 in. deep so they are below the frost line. Other buried cables should be enclosed in sturdy polyurethane or PVC pipes. The pipes should have an inside diameter several times the outside diameter of the cable to protect against earth movements. An excess length of cable in the pipe prevents tensile loads being placed on the cable.

AERIAL INSTALLATION

Aerial installation includes stringing cables between telephone poles or along power lines. Unlike copper cables, fiber-optic cables run along power lines with no danger of inductive interference.

Aerial cables must be able to withstand the forces of high winds, storms, ice loading, and so forth. Self-supporting aerial cables can be strung directly from pole to pole. Other cables must be lashed to a high-strength steel wire, which provides the necessary support. The use of a separate support structure is the usual preferred method.

INDOOR INSTALLATION

Most indoor cables must be placed in conduits or trays. Since standard fiber-optic cables are electrically nonconductive, they may be placed in the same ducts as high-voltage cables without the special insulation required by copper wire. Many cables cannot, however, be placed inside air conditioning or ventilation ducts for the same reason that PVC-insulated wire should not be placed in these areas: A fire inside these ducts could cause the outer jacket to burn and produce toxic gases.

Plenum cables, however, can be placed in any plenum area within a building without special restrictions. The material used in these cables does not produce toxic fumes.

TRAY AND DUCT INSTALLATIONS

The first mechanical property of the cable that must be considered in planning an installation is the outside diameter of the cable itself and the connectors. Here we discuss the cables from Figure 14-1 terminated with SMA connectors. The connector's outside diameter is 0.38 in. (9.7 mm). The outside diameter of the simplex cable is 0.189 in (4.8 mm). The duplex cable has an oval cross section, 0.193 × 0.335 in. (4.9 × 8.5 mm). If the cable must be pulled through a conduit or duct, the minimum cross-sectional area required is 0.79 × 0.43 in. (20 mm × 11 mm) for duplex cable. For simplex cable, the minimum cross-sectional area is determined by the pulling grip used.

The primary consideration in selecting a route for fiber-optic cable through trays and ducts is to avoid potential cutting edges and sharp bends. Areas where particular caution must be taken are corners and exit slots in sides of trays (Figure 14-2).

If a fiber-optic cable is in the same tray or duct with very large, heavy electrical cables, care must be taken to avoid placing excessive crushing forces on the fiber-optic cable, particularly where the heavy cables cross over the fiber-optic cable (Figure 14-3). In general, cables in trays and ducts are not subjected to tensile forces; however, it must be kept in mind that in long vertical runs, the weight of the cable itself will create a tensile load of approximately 0.16 lb/ft (0.25 N/m) for simplex cable and 0.27 lb/ft (0.44 N/m) for

FIGURE 14-2
Corners and exit slots (Courtesy of Canoga Data Systems)

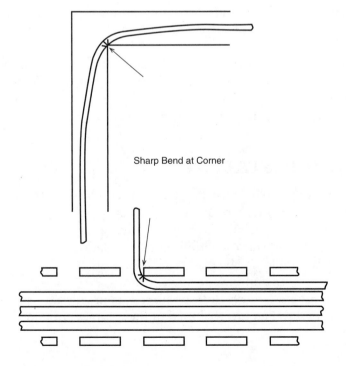

Sharp Bend at Corner

FIGURE 14-3
Crossovers
(Courtesy of
Canoga Data
Systems)

duplex cable. This tensile load must be considered when determining the minimum bend radius at the top of the vertical run. Long vertical runs should be clamped at intermediate points (preferably every 1 to 2 m) to prevent excessive tensile loading on the cable. The absolute maximum distance between clamping points is 330 ft (100 m) for duplex cable and 690 ft (210 m) for simplex cable. Clamping force should be no more than is necessary to prevent the possibility of slippage, and it is best determined experimentally since it is highly dependent on the type of clamping material used and the presence of surface cont-aminants, both on the clamp and on the jacket of the optical cable. Even so, the clamping force must not exceed 57 lb/in. (100 N/cm) and must be applied uniformly across the full-width of the cable. The clamping force should be applied over as long a length of the cable as practical, and the clamping surfaces should be made of a soft material, such as rub-ber or plastic.

Tensile load during vertical installation is reduced by beginning at the top and running the cable down.

CONDUIT INSTALLATIONS

Fibers are pulled through conduits by a wire or synthetic rope attached to the cable. Any pulling forces must be applied to the cable strength members and not to the fiber. For cables without connectors, pull wire can be tied to Kevlar strength members, or a pulling grip can be taped to the cable jacket or sheath. More care, as discussed later in this chapter, must be exercised with cables with connectors.

The first factor that must be considered in determining the suitability of a conduit for a fiber-optic cable is the clearance between the walls of the conduit and other cables that may be present. Sufficient clearance must be available to allow the fiber-optic cable to be pulled through without excessive friction or binding, since the maximum pulling force that can be used is 90 lb (400 N). Since minimum bend radius increases with increasing pulling force, bends in the conduit itself and any fittings through which the cable must be pulled should not require the cable to make a bend with a radius of less than 5.9 in. (150 mm). Fittings, in particular, should be checked carefully to ensure that they will not cause the cable to make sharp bends or be pressed against corners. If the conduit must make a 90° turn, a fitting, such as shown in Figure 14-4, must be used to allow the cable to be pulled in a straight line and to avoid sharp bends in the cable.

FIGURE 14-4
Turn fitting
(Courtesy of
Canoga Data
Systems)

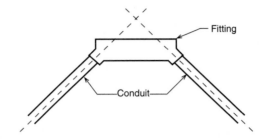

A pullbox is an access point in a conduit. Pullboxes should be used on straight runs at intervals of 250 to 300 ft to reduce the length of cable that must be pulled at any one time, thus reducing the pulling force. Also, pullboxes should be located in any area where the conduit makes several bends that total more than 180°. To guarantee that the cable will not be bent too tightly while pulling the slack into the pullbox, we must use a pullbox with an opening of a length equal to at least four times the minimum bend radius (4.75 in. or 120 mm). See Figure 14-5, which shows the shape of the cable as the last of the slack is pulled into the box. The tensile loading effect of vertical runs discussed in the section on tray and duct installations is also applicable to conduit installations. Since it is more difficult to properly clamp fiber-optic cables in a conduit than in a duct or tray, long vertical runs should be avoided, if possible. If a clamp is required, the best type is a fitting that grips the cable in a rubber ring. Also, the tensile load caused by the weight of the cable must be considered along with the pulling force to determine the maximum total tensile load being applied to the cable.

PULLING FIBER-OPTIC CABLES

Fiber-optic cables are pulled by using many of the same tools and techniques that are used in pulling wire cables. Departures from standard methods are due to several facts: The connectors are usually preinstalled on the cable, smaller pulling forces are allowed, and there are minimum bend radius requirements.

FIGURE 14-5
Pullbox dimen-
sions (Courtesy
of Canoga Data
Systems)

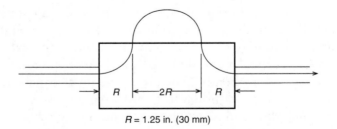

The pull tape must be attached to the optical cable in such a way that the pulling forces are applied to the strength members of the cable (primarily the outer Kevlar layer), and the connectors are protected from damage. The recommended method of attaching a pulling tape to a simplex cable is the "Chinese finger trap" cable grip (Figure 14-6). The connector should be wrapped in a thin layer of foam rubber and inserted in a stiff plastic sleeve for protection. Since the smallest pulling grip available (Kellems 0333-02-044) is designed for 0.25-in. (6.4-mm)-diameter cable and the outside of the simplex cable is only 0.188 in. (4.8 mm), the cable grip should be stretched tightly and then wrapped tightly with electrical tape to provide a firm grip on the cable.

The duplex cable is supplied with Kevlar strength members extending beyond the outer jacket to provide a means of attaching the pulling tape (Figure 14-7). The Kevlar layer is epoxied to the outer jacket and inner layers to prevent inducing twisting forces while the cable is being pulled, since the Kevlar is wrapped around the inner jackets in a helical pattern. The free ends of the Kevlar fibers are inserted into a loop at the end of the pulling tape and then epoxied back to themselves. The connectors are protected by foam rubber and heat-shrink sleeving. The heat-shrink sleeving is clamped in front of the steel ring in the pulling tape to prevent pushing the connectors back toward the rest of the cable.

During pulling of the cable, pulling force should be constantly monitored by a mechanical gauge. If any increase in pulling force is noticed, pulling should immediately cease, and the cause of the increase should be determined. If the pulling wire is subject to friction, the tensile force on the pulling wire will be more than the force applied to the fiber-optic cable, resulting in false readings.

Pull tension can be monitored by a running line tensiometer or a dynamometer and pulley arrangement. If a power winch is used to assist the pulling, a power capstan with adjustable slip clutch is recommended. The clutch, set for the maximum loading, will disengage if the set load is reached. Relying on the experience of those pulling the cable is another alternative.

If necessary during pulling-in, the cable should be continuously lubricated. The lubricant can be poured or brushed on the cable as it enters the duct or conduit. Alternatively, a slit lubrication bag can be pulled ahead of the cable so that the lubricant spills out during the pull. Lubrication should be used only for difficult pulls.

FIGURE 14-6
Simplex pulling grip (Courtesy of Canoga Data Systems)

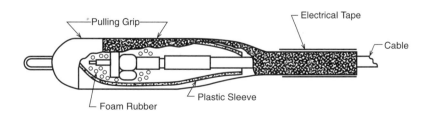

FIGURE 14-7
Duplex pulling
grip (Courtesy
of Canoga Data
Systems)

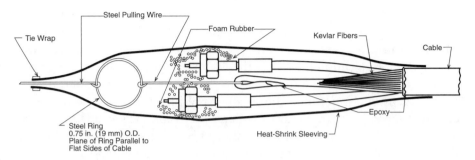

At points such as pullboxes and manholes, where the cable enters the conduit at an angle, a pulley or wheel should be used to ensure that the cable does not scrape against the end of the conduit or make sharp bends. If the portion of the cable passing over the wheel is under tension, the wheel should be at least 12 in. (300 mm) in diameter. If the cable is not under tension, the wheel should be at least 4 in. (100 mm) in diameter.

As the cable emerges from intermediate-point pullboxes, it should be coiled in a figure-8 pattern with loops at least 1 ft in diameter for simplex and duplex cable and proportionally larger diameters for larger cables. The figure-8 prevents tangling or kinking of the cable. When all the cable is coiled and the next pull is to be started, the figure-8 coil can be turned over and the cable paid out from the top. This will eliminate twisting of the cable. The amount of cable that has to be pulled at a pullbox can be reduced by starting the pull at a pullbox as close as possible to the center of the run. Cable can then be pulled from one spool at one end of the run; then the remainder of the cable can be unspooled and coiled in a figure-8 pattern and pulled to the other end of the run.

Bends in the pull should be near the beginning of the pull. Pull forces are lower if the conduit bend is near the beginning. Bends tend to multiply, rather than add, tension. For example, if a cable goes in at 20 lb, it may come out of the bend at 30 lb. If it goes in at 200 lb, it may come out at 300 lb. In the first instance, the bend added 10 lb; in the second it added 100 lb.

SPLICE CLOSURES/ORGANIZERS

A *splice closure* (Figure 14-8) is a standard piece of hardware in the telephone industry for protecting cable splices. You can see them along nearly every aerial telephone run, and they are also used in underground applications. Splices are protected mechanically and environmentally within the sealed closure. The body of the closure serves to join the outer sheaths of the two cables being joined.

A standard universal telephone closure can also be used to house spliced fiber-optic cable. An *organizer panel* holds the splices. Most organizers contain provisions for securing

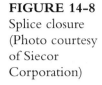

FIGURE 14-8
Splice closure
(Photo courtesy
of Siecor
Corporation)

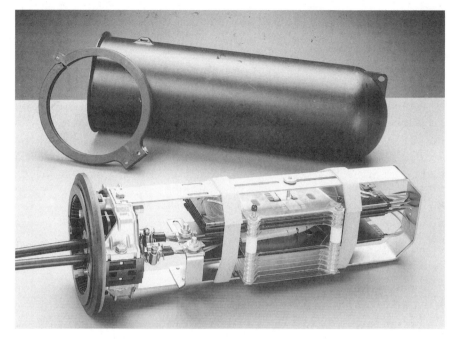

cable strength members and for routing and securing fibers. Metal strength members may be grounded through the closure.

Organizers are specific to the splice they are designed to hold. In a typical application, the cable is removed to expose the fibers at the point they enter the closure. The length of fiber exposed is sufficient to loop it one or more times around the organizer. Such routing provides extra fiber for resplicing or rearrangement of splices.

Closures can hold one or more organizers to accommodate 12 to 144 splices.

Closures have two main applications. The first is a straight splicing of two cables when an installation span is longer than can be accommodated by a single cable. The second is to switch between types of cables for various reasons. For example, a 48-fiber cable can be brought into one end of the closure. Three 12-fiber cables, all going to different locations, can be spliced to the first cable and brought out the other end of the closure.

Management Hardware

Neatness counts. Management hardware is the means by which fiber-optic cables are routed, interconnected, and organized in a neat and manageable way. Once there are more than a few cables in an installation, unless they are managed neatly and efficiently you will end up with a rat's nest of cabling, a tangled mess, and a hopeless mass of confusion. Management hardware allows fibers to be organized so they can be installed correctly and maintained efficiently. It is quite common for cables to be re-arranged as needs change. If you cannot identify the cables and where they are routed, you cannot properly maintain an installation.

Figure 14-9 shows examples of a fiber-optic installation in both an unmanaged and managed state. In the unmanaged state, cables are run directly between the input and output points, with no thought given to keeping things neat and tidy. Cables cannot be easily traced for identification, changing a cable might involve untangling cables, and the likelihood of error is high. In short, pity the poor technician responsible for maintaining the system. The managed version is obviously more coherent, organized, and installed with forethought and care. The payoff is an installation that is easy to manage and maintain.

While all installations have management hardware to some degree, in the successful system the hardware is properly installed and used.

PATCH PANELS

Patch panels are rather straightforward—panels that contains coupling adapters to allow cables to be plugged or unplugged. Most patch panels permit plugging on both sides to achieve maximum flexibility in arranging and rearranging the cables. Others provide cable management to keep the routing and placement of cable orderly. In practice, one side of the patch panel is "fixed" cable that is not intended to be rearranged. An example

FIGURE 14-9 (PART 1)
Neatness counts: Management hardware can be the difference between a mess . . .

FIGURE 14-9 (PART 2)

. . . and a tidy, manageable installation (Photos courtesy of Siecor Corporation)

would be the behind-the-wall cables in a building network. These would run to the rear of the patch panel. Patch cable would then run between the patch panel and the equipment. By rearranging the patch cables, you can connect ports on the equipment to different cables in the building. Another use for a patch panel is to form a transition point between multifiber cables and duplex cables. A 72-fiber cable can connect to 72 separate adapters in the patch panel. Individual or duplex cables then connect to the front of the patch panel. Figure 14-10 shows a patch panel.

DISTRIBUTION/MANAGEMENT HARDWARE

Distribution units and enclosures are similar to patch panels, except they provide additional space to organize fiber. It's always a good idea to leave some extra fiber—this slack is then available if it becomes necessary to reterminate a connection. The distribution units provide an area to store the extra fiber in a large loop.

Distribution hardware comes in both rack-mount and wall-mount configurations, and in a variety of sizes to fit different applications needs. Large units will accommodate hundreds or thousand of fibers in large buildings or telecommunications centers, while smaller units meet more modest needs. Among features found in organizers are the following (the exact mix of features can vary with the specific hardware):

FIGURE 14-10
Patch panel
(Photo courtesy
of AMP
Incorporated)

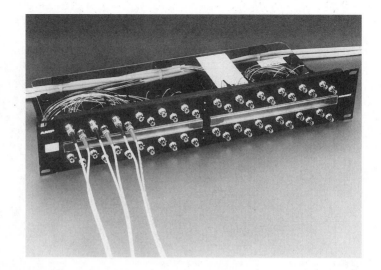

- *Fiber storage:* The unit should allow storage for excess fiber.
- *Cable management features:* Provisions should be made for keeping loops neat and for securing the fiber or cable with tie wraps or other fasteners. Many enclosures have studs or other positioning features that the cable can be wrapped around neatly.
- *Fiber access:* The enclosure should allow multiple points of entry for fiber so that routing from outside to inside is convenient and flexible. The access points should guard against abrasion to the fiber jacket.
- *Protection:* Many units are fully enclosed to protect the fiber.
- *Human access:* How easy is it to gain access to the inside of the box for maintenance or other needs? Front and rear access is handy. Some hardware feature swing-out or pull-out access to patch panels or splices.
- *Security:* Is there a need to lock the box to prevent unauthorized access?
- *External cable routing and management:* Does the hardware support rings or other brackets to help route the cable? Again, such devices are important to keeping the installation neat and maintainable.
- *Labels:* The enclosure should allow all ports to be easily and clearly labeled. All cables and ports must be clearly identified so that the network can be documented. If nobody knows what connects to what, how is the system to be maintained?
- *Multimedia support:* Many installations must support both copper and fiber. Management hardware is available that allows both to be used.

Most management hardware mounts in a 19- or 23-inch-wide rack—standard widths in the telecommunications and electronics industry. Figure 14-11 shows a typical enclosure.

FIGURE 14-11
Management
distribution
hardware (Photo
Courtesy
of Siecor
Corporation)

OUTLETS

You're familiar with the telephone jack in a home or office, used to plug in your telephone (or computer modem). Outlets are similarly available for fiber. Some, such as the one shown in Figure 14-12, provide a connection for both copper and fiber. In an office, there is often a need for many connections: telephone(s), fax, computer network, and so forth. The wall outlet provides the transition between the building cable and the office

FIGURE 14-12
Outlets allow
users to plug into
fiber (or copper)
systems (Photo
courtesy of AMP
Incorporated)

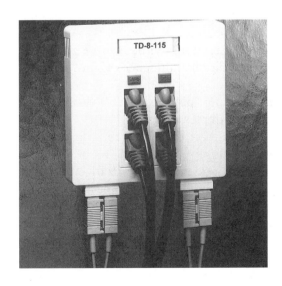

cabling. Interfaces are available in almost every conceivable style. Almost every variation of outlet devised for copper telephone/network cabling has also been adapted to accept fiber connections as well.

SUMMARY

- The two most important concerns in installing a fiber-optic cable are minimum bend radius and maximum tensile load.
- Cables can be buried, strung aerially, or placed in trays and conduits.
- Closure/organizers are used to organize and protect spliced cable.
- Distribution boxes provide protection and organization for fibers and allow versatile and flexible distribution of fibers as application demands.
- Patch panels allow pluggable, rearrangeable interconnection of fibers.
- Wall outlets are similar to electrical outlets in that they allow equipment to be plugged into building wiring.

 Review Questions

1. What are the two most important factors to be considered when installing a cable?
2. Will a cable's minimum allowable bend radius be less during the installation or after?
3. Will a cable's maximum tensile load be greater during the installation or after?
4. Name three additional influences against which a buried cable must be protected.
5. In an intrabuilding application, what distinguishes an ordinary duplex cable from a plenum duplex cable?
6. When pulling a cable through a conduit, to what part must the pulling forces be applied?
7. Why is a fiber-optic rack either 19 in. or 23 in. wide?
8. What type of hardware allows connection between intrabuilding cable and drop cables from equipment?
9. What distinguishes a patch panel from a distribution box for splices?
10. If the total length of an application is twice the distance that can be achieved in a single pull, where is the best place to begin?

chapter fifteen

FIBER-OPTIC SYSTEMS
network and premises applications

LANS

A local-area network (LAN) is a sophisticated arrangement of hardware and software that allows stations to be interconnected and pass information between them. Most business computers are connected to a LAN and home LANs are becoming a topic of interest as many families have more than one computer. LANs allow users to access information, send and receive e-mail, pass files back and forth, and share resources, such as printers, fax machines, and Internet access. In general, a LAN is a network of limited geographical area, such an office, building, or group of buildings. Individuals can also be connected over a wide-area network (WAN). A large company with offices spread all over the country or even the world can connect a number of LANs through WAN links. Most WAN connections, however, are through the public telecommunications system, so we will concentrate here on LANs.

The *topology* of a LAN refers to its physical and logical arrangement. Figure 15-1 shows common network topologies.

- A *bus* structure has the transmission medium as a central line from which each node is tapped. Messages flow in either direction on the bus. Most bus-structured LANs use coaxial cable as the bus.
- A *ring* structure has each node connected serially in a closed loop. Messages flow from one node to the next in one direction around the ring.
- A *star* topology has all nodes connected at a central point, through which all messages must pass.

FIGURE 15-1
Network
topologies

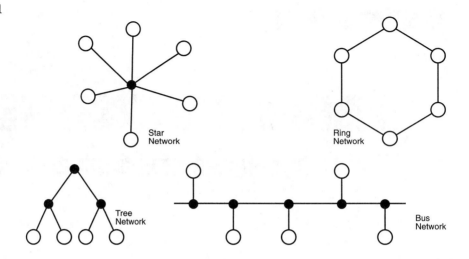

- A *tree* topology uses a branching structure. Most often this forms a *hierarchical star* layout, as used in premises cabling. There are several star points, with each star center feeding a star higher up the chain. A single star occupies the top. Many practical configurations—including networks and premises cabling—use a tree topology.
- Hybrid topologies combine one or more of the basic topologies.

It is important to distinguish between a *physical* topology and a *logical* topology. A logical topology defines how the software thinks the network is structured—how the network is "philosophically" constructed. A physical topology defines how the LAN is physically built and interconnected. For example, the Ethernet LAN is a bus-structured network logically. Its physical topology depends on the specific variety of network. Sometimes it is wired like a traditional bus, which reflects its origins as a bus network running on coaxial cable. More often today it is wired as a star or a hybrid configuration. Regardless of how it is wired or interconnected, the LAN appears to the control software as a bus network. Similarly, although Token Ring is a ring network, it is also configured as a star with all stations connecting through a hub or switch. One reason for the change is that it is increasingly easier and more economical to build intelligence into the hub of the star. Microprocessors, memory, and application-specific chips are less expensive and allow a great deal of sophistication to be incorporated into the hub or switch. The resulting networks tend to be more robust and reliable, and many of the most recent advances (such as switching) takes advantage of the star structure.

Star-wired networks are popular because they increase reliability. Cable breaks in ring and bus structures can bring down the entire network. In a star network, only the affected portion crashes; the rest of the network can continue. The cabling topology recommended in modern LANs and premises cabling is a hierarchical star configuration.

NETWORK LAYERS

As data communications becomes an important part of business, a need arises for universality in exchanging information within and between networks. In other words, there is a need for standardization, a structured approach to defining a network, its architecture, and the relationships between the functions.

In 1978, the International Standards Organization (ISO) issued a recommendation aimed at establishing greater conformity in the design of networks. The recommendation set forth the seven-layer model for a network architecture shown in Figure 15-2. This structure, known as the Open Systems Interconnection (OSI), provides a model for a common set of rules defining how parts of a network interact and exchange information. Each layer provides a specific set of services or functions to the overall network. The three lower layers involve data transmission and routing while the top three layers focus on user needs. The middle layer—transport—provides an interface between the lower three layers and the top three layers. Unfortunately, each layer has numerous standards that define the layer differently. The network systems in this chapter, for example, define the lower layers differently.

The following is a brief description of each layer.

Physical Layer

The physical layer is the most basic, concerned with getting data from one point to another. This layer is comprised of the basic electrical and mechanical aspects of interfacing to the transmission medium. This includes the cable, connectors, transmitters, receivers,

FIGURE 15-2
OSI seven-layer
network model

Internetworking Device	OSI Model Layer		Layer Function
	7	**Application**	Specialized functions, such as file transfer, terminal emulation or electronic mail
	6	**Presentation**	Data formatting and character code conversion
	5	**Session**	Negotiation and establishment of a connection with another node
↑**Gateway**	4	**Transport**	Provision for end-to-end delivery
Router	3	**Network**	Routing of information across multiple networks
Bridge	2	**Data Link**	Transfer of units of information, sequencing, and error checking
Repeater	1	**Physical**	Transmission of raw data over a communication channel

and signaling techniques. Most of this book is concerned with the physical layer. Each specific type (or subtype) of network defines the physical layer differently.

While not used in LANs, the most widely used physical layer is RS-232, which defines the voltage levels of signals, the function of each wire, the type of connector, and so forth. The com or serial port of a personal computer implements an RS-232 physical layer. You can see that each specific need requires a different definition for this layer. Simply changing the cable type or connector creates a different layer.

Data-Link Layer

The data-link layer provides reliable transfer of data across the physical link. This layer establishes the protocols or rules for transferring data across the physical layer. It puts strings of characters together into messages according to specific rules, manages access to and from the physical link, and ensures the proper sequence of transmitted data.

Networks like Ethernet and Token Ring exist on the physical and data-link layers. In a PC network, chips on the PC's network interface card perform the physical and data link functions. Software in the computer performs the functions of the higher layers. From the viewpoint of this book, building cabling and networks are distinguished at the physical and data-link layers.

Network Level

The network control layer addresses messages to determine their correct destination, routes messages, and controls the flow of messages between nodes. The network level, in a sense, forms an interface between the PC and the network. It controls where the data on the network goes and how it is sent. The Internet uses protocols called transmission control protocol and internet protocols (TCP/IP) to control communications between computers on the net. IP works on the network layer.

Transport Level

This layer provides end-to-end control once the transmission path is established. It allows exchange of information independent of the systems communicating or their location in the network. This level performs quality control, checking that the data is in the right format and order. TCP exists on the transport layer.

Session Layer

The session-control layer controls system-dependent aspects of communication between specific nodes. It allows two applications to communicate across the network.

Presentation Layer

At the presentation layer, the effects of the layer begin to be apparent to the network user. This layer translates encoded data into a form for display on the computer screen or for printing. In other words, it formats data and converts characters. For example, most computers use the American Standard Code for Information Interchange (ASCII) format to represent characters. Some IBM equipment uses a different format, the Extended Binary Coded Decimal Information Code (EBCDIC). The presentation layer performs the translation between these two formats. Microsoft Windows performs presentation layer functions.

Application Layer

At the top of the OSI model is the application layer, which provides services directly to the user. Examples include resource sharing of printer or storage devices, network management, and file transfers. Electronic mail is a common application-layer program. This is the layer directly controlled by the user.

In practice, networks work from the top layer of one station (the message originator) to the top of another station (the message recipient). A message, such as electronic mail, is created in the top presentation layer of one workstation. The message works its way down through the layers until it is placed on the transmission medium by the physical layer. At the other end, the message is received by the physical layer and travels upward to the presentation level. The electronic mail is read at the presentation level.

This layered approach to building a network holds two benefits. First, an open and standardized system permits equipment from different vendors to work together. Second, it simplifies network design, especially when extending, enhancing, or modifying a network. For example, while most LANs use copper wire, either coaxial cable or twisted pairs, adding a fiber-optic point-to-point link involves only the physical layer. Higher levels of the OSI model do not care how the physical layer is actually implemented; they care only that the physical layer follows certain rules in interacting with higher levels.

Bear in mind that the OSI model is a reference model. While the ISO has created protocol standards for each level, they are not widely used. Different systems use different protocols. For example, Novell NetWare and Microsoft Windows Networking use different protocols for the different layers. IBM's System Network Architecture (SNA) and Digital Equipment's Network Architecture (DNA) have yet different rules. Nor does a system have to use all seven layers; a system can combine two layers into a single set of protocols.

Since Novell NetWare is a popular network operating system for PCs, here, by way of example, are the protocols used:

7. Application: NetWare shell
6. Presentation: Network Core Protocols (NCP)
5. Session: NetBIOS Emulator
4. Transport: Sequenced Packet Exchange (SPX)
3. Network: Internet Packet Exchange (IPX)
2. Data Link: Open Data-Link Interface (ODI)
1. Physical   (Specific network type: Ethernet, Token Ring, etc.)

A network can be connected to other networks of the same or different type. Sometimes users break a large network into several smaller segments to make the network more efficient.

Figure 15-2 also shows the devices required to interconnect either different networks or segments of the same network. The simplest device is a repeater operating at the physical layer. The repeater simply takes an attenuated signal, amplifies it, retimes it, and sends it on its way. It is a "dumb" device since it does not look at the message; it only regenerates the pulses.

A bridge operates at the next level, the data-link level, to link different networks that use the same protocol. The bridge differs from a repeater in its built-in intelligence. It passes from one segment to another messages intended for the second segment. It controls which traffic passages through it and which remains local to the originating segment. The bridge examines the message and determines its destination. Bridging can help improve network efficiency by reducing the number of stations that a message must pass through.

A router operates at the network-control level and can handle different protocols. It offers greater sophistication than a bridge. First, it can handle different protocols so you can hook together networks using unlike protocols. Second, it can add additional information to a frame to allow routing over larger networks such as an X.25 packet-switching network used for long-distance transmissions over the public telephone system. Third, it can choose the best route if several routes are available. Many routers also compress the data to make routing, especially over slower-speed long-distance lines, more economical. For example, a router with 4:1 compression can reduce a 8-megabyte file to a mere 2 megabytes for transmission.

Bridges and routers can be used locally or remotely. In a local application, a single bridge or router connects each LAN segment. In a remote application, typically over the public telephone network or private or leased lines, a bridge or router is required at each end. A gateway works at higher levels, serving as an entry point to a local area network from a larger information resource such as a mainframe computer or a telephone network.

The ISO model is a handy framework for understanding the structure of a network. Some network models do not even have seven layers.

ACCESS METHOD

Access refers to the method by which a station gains control of the network to send messages. Two methods are carrier sense multiple access with collision detection (CSMA/CD) and token passing.

In CSMA/CD, each station has equal access to the network at any time (multiple access). Before seizing control and transmitting, a station first listens to the network to sense if another workstation is transmitting (carrier sense). If the station senses another message on the network, it does not gain access. It waits awhile and listens again for an idle network.

The possibility exists that two stations will listen and sense an idle network at the same time. Each will place its message on the network, where the messages will collide and become garbled. Therefore, collision detection is necessary. Once a collision is detected, the detecting station broadcasts a collision or jam signal to alert other stations that a collision has occurred. The stations will then wait a short, random period for the collision to clear and begin again.

In the token-passing network, a special message called a *token* is passed from node to node around the network. Only when it possesses the token is a node allowed to transmit.

FRAMES

The information on a network is organized in frames (also called packets). A frame includes not only the raw data, but a series of framing bytes necessary for transmission of the data. Figure 15-3 shows the frame formats for both the Ethernet, FDDI, and ATM data transmissions. Notice that ATM has a short, fixed length format of 53 bytes. We'll discuss the significance of this later in this chapter when we describe ATM. Other frames can

FIGURE 15-3
Frame formats:
Ethernet, FDDI,
and ATM

Preamble	Start of Frame	Destination Address	Source Address	Length	Data	CRC
7 bytes	1 byte	6 bytes	6 bytes	2 bytes	46 - 1500 bytes	4 bytes

IEEE 802.3 Frame

Preamble	Start of Frame	Packet ID	Destination Address	Source Address	Data	CRC	End of Frame + Status
≥ 7 bytes	1 byte	1 byte	2 or 6 bytes	2 or 6 bytes	0 - 4486 bytes	4 bytes	≥ 2 bytes

FDDI Data Frame

Header	Data
5 bytes	48 bytes

ATM Cell

also be used. For example, besides the data frame, a token-ring network also uses a token frame for passing the token around the network.

Here is a brief description of the elements of an FDDI data frame.

The *preamble* indicates the beginning of a transmission. The preamble is a series of alternating 1s and 0s—1010101010 . . . —that allows the receiving stations to synchronize with the timing of the transmission.

The *start of frame* is a special signal pattern of 10101011 that signals the start of information.

The *packet ID* identifies the type of packet, such as data, token, and so forth.

The *destination address* is the address of one or more stations that are to receive a message. In a network, each station has a unique identifier known as its address.

The *source address* identifies the station initiating the transmission.

The *data* is the information, the point of the transmission.

The *CRC* or *cyclic redundancy check* is a mathematical method for checking for errors in the transmission. When the source sends the data, it builds the CRC number based on the patterns of the data. The receiver does the same thing as it receives the data. The receiving station also builds a CRC. If the receiver CRC matches the transmitted CRC, no errors have occurred. If they don't match, an error is assumed, the transmitter is informed, and the transmission is sent again.

The *end of frame* informs the receiving station that the transmission is over and also contains the status of the transmission. The receiving station marks the status to acknowledge receipt of the packet.

SHARED VERSUS SWITCH MEDIA

Early networks like Ethernet and Token Ring use a shared-media concept in passing messages. The message goes to every station on the network, each station checking the address to see if the message belongs there.

More recently, switched media has become popular. In a switched-media network, a central device such as a hub contains a switching network. The hub checks the distinction address and then sends the message only to the correct station. In this case, the hub is called a *switch*.

Switched-media networks rely on a star-wired configuration, since there must be intelligence at the star's center to decode the message and perform the switching. Network topologies like rings and buses must be shared media since the intelligence for decoding messages lies in the stations. The message must go to each station.

Figure 15-4 shows the difference between shared and switched media networks. Switched networks make much more efficient use of network resources. The time a message spends on the network is reduced significantly since it passes directly from the originating station through the hub to the destination station. Switching increases network availability and can serve as an alternative to a higher speed network. A 10-Mbps switched

FIGURE 15-4
Shared versus
switched media

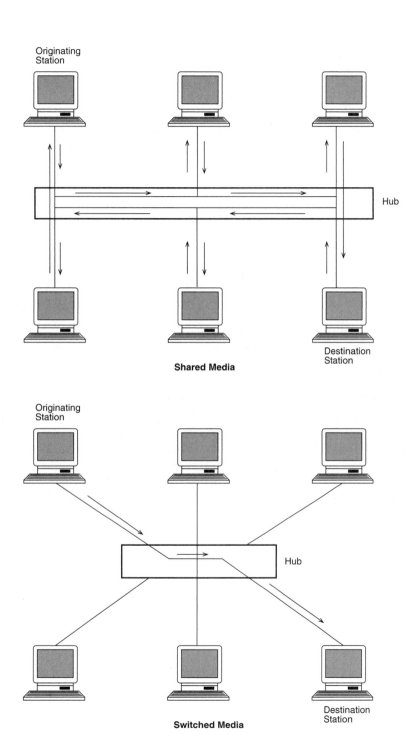

Shared Media

Switched Media

Ethernet network (rather than a 100-Mbps Fast Ethernet) may serve as a cost-effective upgrade to a 10-Mbps shared-media Ethernet.

The key difference between shared- and switched-media networks lies in the bandwidth available to each user. In a shared-media Ethernet network, the 10-Mbps bandwidth is shared by the network and attached users. As the number of stations increases, the bandwidth per station drops. In a switched Ethernet network, each station has its own 10-Mbps link. The aggregate or total bandwidth is determined by the central switch, which has a bandwidth many times greater than 10 Mbps.

QUALITY OF SERVICE

Quality of service (QoS) refers to the ability of a network to deliver an application with the quality it requires. QoS became of interest as the delivery of video and other time-sensitive, real-time applications became practical. High-quality video requires a steady stream of data. To avoid jerky, awkward video, the network must be able to guarantee a certain level of delivery. The original networks were incapable of such guarantees. Because the bandwidth was shared between many users who traded large and small files, a steady delivery of data was impractical. It wasn't so much that the network did not have the bandwidth: The bandwidth was there, but there was no mechanism for allocating it among users.

By the early 1990s, newer networks had built-in priority schemes to allow priority to be assigned to real-time data. Asynchronous transfer mode (ATM), a network we will describe later in this chapter, offered a particularly rich approach to QoS, but other networks are attempting to build QoS into their systems.

ETHERNET

Ethernet is a CSMA/CD LAN available in a wide variety of subtypes or "flavors" ranging in data rates from 10 Mbps to 1000 Gbps. In terminology, Ethernet flavors are distinguished as xBASEy. The x refers to the speed. The y refers to the transmission medium and, in the case of fiber optics, to the type of source used. Thus 10BASE-T is a 10-Mbps network running over unshielded twisted-pair cable. 100BASE-FX is a 100-Mbps network running over fiber-optic cable. The *BASE* refers to the fact that most Ethernet flavors use baseband signalling. There is a broadband version, which is not widely used and is not discussed here.

Ethernet comes in three main speeds: original Ethernet at 10 Mbps, Fast Ethernet at 100 Mbps, and Gigabit Ethernet at 1 Gbps. Each type supports both copper and fiber versions. Figure 15-5 summarizes many of the different types of Ethernet.

FIGURE 15-5
Flavors of
Ethernet

Flavor		Cable Type	Segment Length*	Speed
Ethernet	10BASE-5	Thick Coax	500 m	10 Mbps
	10BASE-2	Thin Coax	185 m	
	10BASE-T	Cat. 3 UTP	100 m	
	10BASE-FP	Fiber	1000 m	
	10BASE-FB	Fiber	2000 m	
	10BASE-FL	Fiber	2000 m	
Fast Ethernet	100BASE-TX	Cat. 5 UTP	100 m	100 Mbps
	100BASE-T4	Cat. 3 UTP	100 m	
	100BASE-FX	Fiber	2000 m	
Gigabit Ethernet	1000BASE-T	Cat. 5 UTP	100 m	1 Gbps
	1000BASE-CX	Shielded Copper	25 m	
	1000BASE-SX (850 nm)	Fiber (62.5)	260 m	
		Fiber (50)	525 m	
	1000BASE-LX (1300 nm)	Fiber (62.5)	550 m	
		Fiber (50)	550 m	
		Fiber (SM)	3000 m	

*Segment length is the backbone cable length for 10BASE-5 and -2. For others, it is the distance from hub to station (or other attached device).

The original Ethernet flavors (10BASE-2 and 10BASE-5) used coaxial cable as a backbone with each station attached to the backbone in a bus network. Coaxial cable was replaced with the advent of 10BASE-T using twisted-pair wiring and hubs to create a star-wired network. A hub or switch-based star wired network is common to all types of Ethernet except 10BASE-2 and -5.

Figure 15-6 shows a typical configuration for an Ethernet network. The figure shows a hierarchical star. Notice that different speeds can be mixed. While it would be nice to provide the latest and greatest Ethernet link to each desktop computer, economics makes this impractical. Gigabit ports are very expensive, while Fast Ethernet and 10BASE-T Ethernet are relatively cheap. A typical configuration will deliver 10 Mbps or 100 Mbps to the desktop and use either Fast or Gigabit Ethernet in the backbone to provide high-speed links between hubs or switches.

TOKEN RING

Token Ring is a token-passing network originated by IBM. It is most popular with so-called IBM shops, companies with a heavy commitment to larger IBM computers. Operating at 4 and 16 Mbps, Token Ring fell behind other high-speed networks.

FIGURE 15-6
Typical Ethernet
network

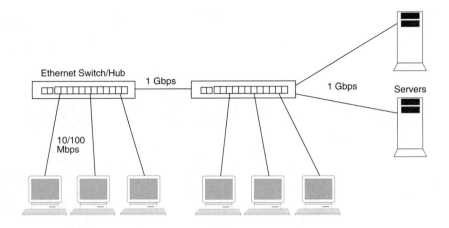

Recently switched Token Ring and 100-Mbps Token Ring have offered a migration path for users who need additional performance.

Like Ethernet, Token Ring is today applied as a star-wired network. If you actually trace the path of the signals through the hub, you would indeed find a ring, with a signal passing from station to station.

FDDI

The fiber distributed data network (FDDI) is also a token-passing network that deserves special mention in the history of fiber optics for three reasons.

First, it was the premier LAN standard designed specifically for fiber optics from the beginning. While copper alternatives came later, FDDI was conceived as a high-speed network taking advantage of optical-fiber technology. Figure 15-7 shows an FDDI network.

Second, it was the first high-speed network operating at 100 Mbps. At a time when Ethernet ran at 10 Mbps and Token Ring at 16 Mbps, FDDI's speed represented a significant step. It was several years before other networks offered equivalent speeds.

Third, FDDI was a robust network capable of handling heavy-duty networks. FDDI operates at 1300 nm and can handle over 1000 stations over a distance of 100 km. FDDI uses hubs (usually called concentrators) to achieve a star-wired arrangement. Typically, stations are connected to hubs and hubs are connected in a ring. Each cabling segment (station-to-hub or hub-to-hub) is essentially a point-to-point link. Several variations of FDDI cabling have evolved. They are typically referred to as *PMDs,* for physical medium dependent. In other words, each PMD defines a physical layer of FDDI.

Multimode Fiber PMD

The MMF-PMD is the original FDDI specification for multimode fiber. It uses the FDDI duplex MIC connector and provides keying for both the port type and polarity. It permits link lengths of 2 km between stations, with an 11 dB loss budget.

The preferred fiber is the well-known 62.5/125μm fiber, although 50/125, 85/125, and 100/140 fibers are also allowed. (The 85/125μm fiber shows the age of the standard. The fiber was once a contender, but has been largely supplanted.)

Low-Cost Fiber PMD

In an effort to lower the cost of FDDI links, LCF-PMD evolved. After looking at several alternatives in reducing costs—including using different fibers (such as 200/230 step-index fiber and plastic fiber)—the committee decided the real cost of the link lay in the transceiver, not in the fiber. One approach—of using less expensive 850-nm devices—was discarded because of the difficulties involved in ensuring each end of the link was the same. The LCF-PMD relaxes the performance requirements for transmitters and receivers and replaces the FDDI connector with a lower cost SC connector. The resulting link length is 500 meters, only 25% of the distance allowed by MMF-PMD but still a considerable length for intrabuilding runs. The allowed link power budget is reduced from 11 dB to 7 dB.

For LCF-PMD, the 62.5/125μm fiber is still preferred. Besides the same alternative fibers allowed by MMF-PMD, the LCF-PMD also allows 200/230 fiber.

Single-Mode Fiber PMD

SMF-PMD covers single-mode fibers. The standard defines two categories of transceivers: Category I transceivers operate at the same power levels as MMF-PMD transceivers, while Category II transceivers are more powerful. The loss budget for a Category II transceiver is 32 dB. Transmission distances are about 40 km for a Category I link and 60 km for a Category II link.

Twisted-Pair PMD

TP-PMD allows the use of copper cable in FDDI links. It permits 100-meter runs between stations.

One attractive feature of FDDI is its dual-ring structure (Figure 15-8). Signals travel around the ring in opposite directions. If a break occurs, the ring wraps to form a smaller ring. This feature increases reliability, which is an important consideration in backbone applications.

FDDI did not gain the market acceptance it perhaps deserved. Part of the reason was cost: FDDI was expensive, especially in its early days before some of the significant cost

FIGURE 15-7
FDDI application
(Illustration cour-
tesy of AMP,
Incorporated)

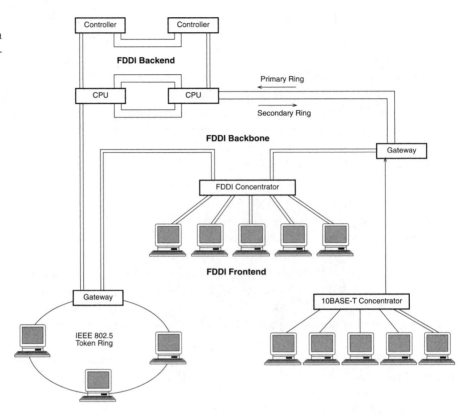

reductions in fiber-optic technology. If FDDI did not bring 100-Mbps performance to the desktop, it was more widely applied in the backbone.

SOME NETWORK DEVICES

We've looked at some of the types of networks and have seen how star-wired systems are at the heart of most networks. Thus hubs and switches become a central player in the network. These devices can range from large enterprise-level devices to workgroup-level devices. The differences are a matter of scale. Enterprise hubs and switches are typically chassis-based systems (see Figure 15-9) that accept plug-in boards. Many enterprise switches or hubs are multiprotocol devices that can handle different types of networks. For example, a switch may contain many plug-in modules for Fast Ethernet links to PCs, one or more gigabit Ethernet links to servers, and even ATM links to the WAN.

Workgroup hubs and switches (Figure 15-10) are more modest, having fewer ports. They are designed for managing smaller networks or subnetworks. Workgroup hubs might

FIGURE 15-8
FDDI's dual-ring structure and ring wrap

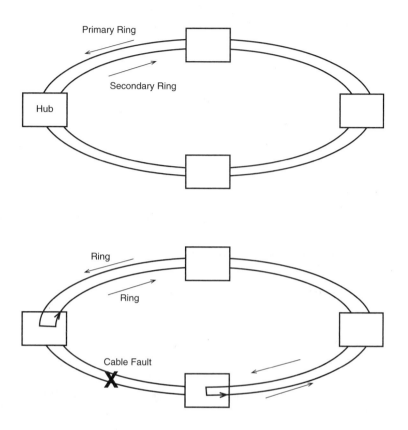

still have installable modules to allow different types of uplinks (such as Fast Ethernet or gigabit Ethernet, copper or fiber). Often the station ports are copper based, while ports used to link switches over greater distances are fiber.

To connect to a network, PC must have a network port. Most often this is provided by a plug-in network interface card (NIC). The NIC in Figure 15-11 accommodates both copper and fiber Ethernet.

FIGURE 15-9
Enterprise switch
(Courtesy of
3Com)

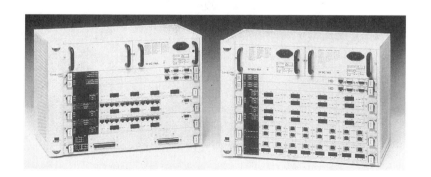

FIGURE 15-10
Workgroup
switch (Courtesy
of 3Com)

One drawback to widespread use of fiber in LANs is the installed base of copper components. The decision between copper and fiber is often presented as an either/or choice. Consider that if companies have hundreds or thousands of PCs with copper NIC cards and lots of hubs with copper ports, there may be some resistance to fiber. In addition, more and more computers are equipped with network ports and some do not use a plug-in card: a copper network port is built directly onto the motherboard. One solution is the media converter, an example of which is shown in Figure 15-12. The media converter performs one simple task: it converts signals between optical and electrical. Plug copper in one and get optical out the other. Media converts can be a relatively inexpensive way to allow both copper and fiber to coexist. For example, you can wire your building with fiber, buy fiber-based hubs and switches, and use PCs with either copper or fiber NICs. Fiber-based PCs can plug in directly to the building wiring. For copper-based PCs, a copper cable from the PC to the media converter saves the investment in PCs.

FIGURE 15-11
NICs (Photo
courtesy of Sun
Conversion
Technologies,
Inc.)

FIGURE 15-12
Media converter
(Photo courtesy
of Sun
Conversion
Technologies,
Inc.)

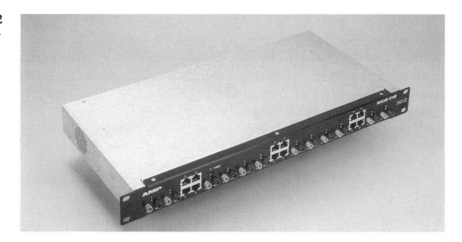

The wider availability of media converters, fiber NICs, and fiber ports on hubs and switches brings greater options in premises cabling systems, our next subject.

PREMISES CABLING

Premises cabling refers to the wiring of buildings for networks, telephones, and other needs. The modern concept of premises cabling is the so-called structured cabling system. The idea is that you can wire your building to a certain level of performance and that any application that fits within those performance parameters will operate. Put another way, the cabling system is application independent. If you install a 100-MHz cabling system, any network or other application that runs at frequencies up to 100 MHz will run on your system.

Two types of cables are preferred for structured premises cabling systems: unshielded twisted-pair (UTP) cable and fiber-optic cable. The unshielded twisted-pair cable consists of pairs of wires twisted together and is "graded" by category. Category 3 cable is rated for applications to 16 MHz and Category 5 is rated for applications to 100 MHz. (There is a Category 4 cable rated to 20 MHz, but few people talk about it and fewer people use it.) In 1998, Category 5 UTP was the standard high-performance copper cable for premises cabling. Category 6 and 7 cables, rated for 250 MHz and 600 MHz, are in the process of being standardized and will find application in the future for high-speed networks. Note that Category 5 cable can support Gigabit Ethernet. This is done in two ways: First, the signal is divided among four pairs of cable, and second, advanced encoding methods ensure that the frequency on each pair is kept under 100 MHz. It is important to remember that you should not confuse the operating frequency in megahertz and the data rate in megabits per second. In other words, MHz ≠ Mbps in many cases.

The "magic" number with UTP is 90 meters. This is the maximum recommended distance for cabling runs.

The preferred fibers for premises cabling are 62.5/125-μm multimode fiber and single-mode fiber. Some applications recommend 50/125-μm multimode fiber and this cable is usually considered an alternative. The 85/125 and 100/140 fibers mentioned for FDDI no longer find widespread use in premises cabling.

A premises cabling application is divided between into two sections: the backbone or vertical cable and the horizontal cable. Backbone cable typically runs between network equipment located on different floors. Horizontal cable runs from the equipment to the work area or office. In addition, the structured system can encompass campus applications in which several buildings are interconnected. There are four basic consolidation points, as shown in Figure 15-13: the main cross connect, the intermediate cross connect, the wiring closet, and the work area. (The intermediate cross connect is optional and is required only in larger applications.) All three are places where equipment is located or where there is a transition between cables.

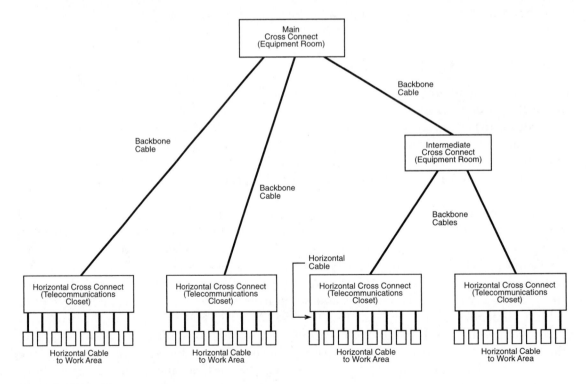

FIGURE 15-13
Main consolidation points in a premises cabling network

Fiber-optic cable is popular in backbone, interfloor applications, but is still making headway in the horizontal run. The application of fiber in horizontal runs is widely called *fiber to the desk*. FTTD has been slowed by several factors:

- *Cost* Fiber systems have traditionally been more expensive than copper counterparts. The cost of fiber has been steadily falling to narrow the gap between fiber and copper. Fiber components will undoubtedly continue to drop—the emergence of the MT-RJ, LC-smell-form-factor connectors, transceivers, and other components are examples. In addition, the cost of installing fiber has become lower as components have improved for speedier, easier application.

- *Familiarity* Copper is familiar to both users and installers. Users were concerned with the supposed fragility of glass. Installers, trained in copper, were also hesitant about fiber. As fiber becomes commonplace, resistance to it is disappearing. Fiber is no longer the brave new world: It is simply another, better transmission medium.

- *Application requirements* The earliest networks, such as 10-Mbps Ethernet operated well over copper cable. Regardless of the recognized reliability of fiber, users did not feel compelled to switch. With higher-speed networks, fiber becomes more attractive as its signal-transmission characteristics become more closely matched to the application.

Figure 15-14 shows the basics of a premises cabling application. The main cross connect contains network equipment (such as hubs or switches) connected through a backbone cable to the telecommunications closet on each floor. This closet typically also contains network equipment. From the telecommunications closet, the horizontal cable runs to an outlet in the work area. A patch cable plugs into the outlet to connect the equipment in the work area to the horizontal cable. In order to provide a fixed "behind-the-wall" structured cabling system, this cabling is terminated at a patch panel or (for fiber) organizer. The equipment is connected to the structured cabling system by a patch cable that plugs into the patch panel.

There are three basic approaches to a structured cabling system within a building: distributed, centralized, and zoned. The differences depend on how and where equipment is located and how the horizontal run is accomplished.

FIGURE 15-14

Aspects of horizontal cable

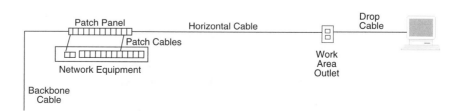

DISTRIBUTED NETWORK

Figure 15-15 shows a distributed system. Here equipment is located in the telecommunications room of each floor and also usually in the main cross connect room. The vertical backbone cable connects the main cross connect room to each telecommunications closet. The backbone cable between floors and the horizontal cable running to offices are clearly separated. The network equipment (hubs and switches) sits between the backbone and horizontal cable. In a typical system, the backbone runs to a patch panel and connects to the equipment through a patch cable. The equipment connects through another patch panel to the horizontal cable. The horizontal cable runs are limited to 90 meters—plus a total of 10 meters for patch cables in the telecommunications room and office.

The advantage of a distributed network is that it is quite compatible with both copper and fiber cabling. You can think of a distributed system as the "classic" way of connecting a network and most networks today use a distributed architecture. The drawback to the distributed system, as you will see, is that it typically is less efficient in its use of equipment than a centralized system. In addition, from the standpoint of fiber, the 90 meters allowed for horizontal cable clearly does not take advantage of the transmission characteristics of fiber. The centralized network does take advantage of them.

FIGURE 15-15
Distributed network (Illustration courtesy of AMP Incorporated)

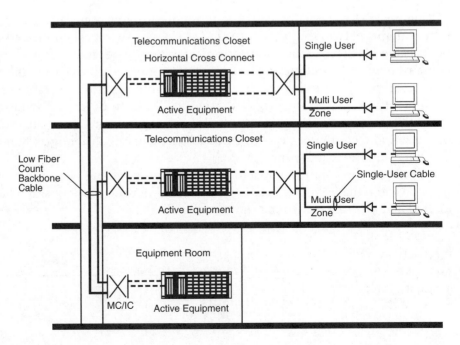

CENTRALIZED NETWORK

Figure 15-16 shows a centralized network. The centralized network takes advantage of the distinct advantages of fiber. The distributed network limits the horizontal cabling to 90 meters. The centralized approach blurs the distinction between the horizontal and vertical cable to allow runs from the main equipment closet to the work area. No equipment is required in the telecommunications closet. The fiber can be spliced to the transition point between vertical and horizontal cable or it can be run directly without splicing. The maximum recommended distance for a multimode fiber is 300 meters.

Centralized networks offer several advantages. They can cut equipment costs by consolidating equipment in a single location. Consider, for example, an application that requires 13 users on each floor. Most network hubs are available with 8, 12, 16, or 24 ports. With a distributed network, you would need a 16-port hub on each floor. For four floors, you need four hubs or a total of 64 ports for only 52 users. Plus, you might want to connect each floor through a switch in the main closet. With a centralized network, you can more closely match the number of ports in your equipment to your actual needs.

Centralized networking can also reduce the time and costs of administering a network. Networks sometimes need maintenance, for example, rearranging cables to accom-

FIGURE 15-16
Centralized
network
(Illustration
courtesy of AMP
Incorporated)

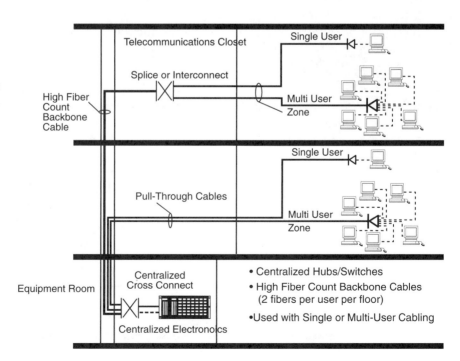

modate users who switch offices, adding new users and removing users who should no longer be connected. Because all equipment and cross connects are in one location, you can perform all moves, adds, and changes from a single location.

While a centralized structured network was designed to take advantage of fiber, it can also be built with copper cable, but only if the total distance does not exceed 90 meters.

There is one potential drawback to the centralized system in Gigabit Ethernet. It has a recommended distance of 260 meters over 62.5/125 μm at 850 nm, which is 40 meters shy of the 300 meter maximum recommended for centralized fiber systems. This drawback can be removed by using 50/125-μm fiber, which has a recommended transmission distance of 525 meters with Gigabit Ethernet, or by transmitting at 1300 nm. There have also been proposals to produce 62.5 μm with a higher bandwidth at 850 nm.

ZONE NETWORK

A zone system (Figure 15-17) is a specialized form of distributed system. Here the horizontal cable is divided into two sections. A "backbone" zone cable is run to a work area. Individual cables are then broken out and run to the work area. In a zone system, an MT 12-fiber array connector can be used from the equipment closet to the zone. A breakout box will divide the ribbon fiber into separate interfaces. The cable on the trunk side is a multifiber cable terminated in an MT connector. Internally, the breakout box will have a small section of fiber, with an MT connector at one end and SC connectors at the other. The box becomes an MT-to-SC interface converter.

FIGURE 15-17
Zone network
(Illustration
courtesy of AMP
Incorporated)

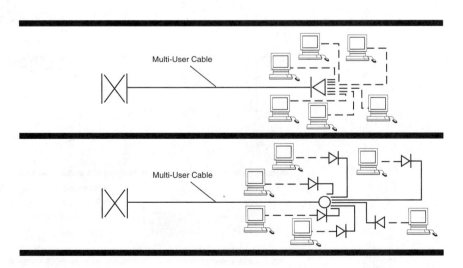

COMPUTER SYSTEMS

Structured cabling systems are also used in large-scale computer applications to bring order to a disorderly mess. A mainframe computer system typically uses underfloor wiring to interconnect the mainframe to its many peripherals using bus and tag cabling. In bus and tag cabling, individual or groups of coaxial cables were run from one device to another. Some of the cables were as thick as your forearm and about as stiff. Each cable was cut to the specific length required for the application. The result was a rat's nest of confusion and mess. Maintaining cables and adding or moving equipment was difficult because of the disorderly way the cables were run. Figure 15-18 contrasts the traditional copper bus and tag configuration with a structured fiber system.

Multifiber array connectors allow a neat, structured approach. Not only are the cables small and flexible, they also allow order to be brought to the cabling. A typical system uses MT array connectors in the under-the-floor trunk and ESCON cable assemblies from a breakout box to the device. This structured approach makes on-going maintenance of the system easier. Adding equipment or moving equipment is easier. Fiber also gives the additional advantage of allowing devices to be remotely located kilometers away.

IBM's ESCON calls its system the Fiber Transport System. A number of vendors offer compatible offerings.

FIGURE 15-18 Copper versus fiber computer system cabling (Courtesy of IBM)

Before (Copper)   After (Fiber)

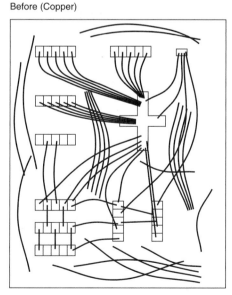
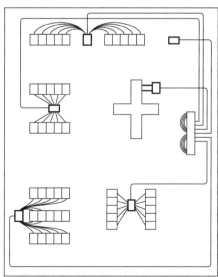

ESCON

As mentioned in Chapter 1, the IBM ESCON system, introduced in 1990, was the first fiber-optic system offered as standard equipment for connecting a mainframe computer to peripheral controllers. Previous fiber-optic links were options, used to extend the channel length, but not to improve the performance of the channel. The channel is the means by which a mainframe computer communicates with peripheral control units. For example, the banks of tape drives in some large computer sites store enormous amounts of information, but are very slow in comparison to the host computer. Even disk drives, which are greatly faster than tape drives, can't keep up with the mainframe. Peripheral control units are specialized computers that manage peripherals such as tape drives, disk drives, and even communications. When the mainframe requests information, it communicates with the peripheral controller, which in turn finds the information on a storage unit and sends it to the mainframe in a quick burst.

Even so, a copper channel is slow and distance-limited. The standard IBM copper-based channel operated a 4.5 Mbps over a maximum distance of 400 feet using the unruly bus-and-tag arrangement discussed above. Transmission is typically in parallel, with each bit of a byte traveling on a separate wire.

ESCON operates at 17 Mbytes/s over distances up to 40 km on single-mode fiber. ESCON works in a point-to-point configuration, where units are directly connected through a fiber-optic link. It also works in a switched environment (similar to switched Ethernet) through a switching device called a Director. A channel can have two Directors in the path. Directors not only add the flexibility of a switched-media arrangement, they also extend transmission distances by up to a factor of three. Transmission distances work out like this:

62.5/125-μm fiber:	3 km point-to-point
	9 km with two Directors
50/125-μm fiber:	2 km point-to-point
	6 km with two Directors
Single-mode fiber:	20 km point-to-point
	26 km with 2 Directors

Each director can have from 8 to 248 ports. Thus ESCON can handle large-computer installations—the data centers with multiple mainframes and hundreds of storage units and other peripherals. The ability to separate devices through single-mode links allows greater flexibility and safety in configuring a system since not all devices must be located in a single place.

Figure 15-19 shows the basics of an ESCON network.

ESCON is combined with the Fiber Transport Services to create a fiber-based channel structure that is faster, neater, and easier to physically manage. The ESCON system has been industry standardized as the Single-Byte Command (SBCON).

FIGURE 15-19
ESCON system

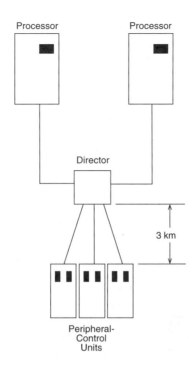

FIBRE CHANNEL

Fibre Channel serves a similar purpose as ESCON. It is designed to connect computer systems and their subsystems (such as disks and controllers) together. Unlike ESCON, it is a much richer and more varied system, offering several topologies, cabling choices, speeds, and distances. In addition it can be extended to perform like a network. Indeed, Fibre Channel is considered to be a *storage area network* (SAN), sharing the same concepts for interconnecting equipment as a LAN.

SANs are a recent concept. While the need for fast, flexible communication between storage units, other peripherals, and servers is apparent, the 90s have seen a surge in demand for storage capacity. Everything is stored on computers today. Applications like data warehousing, on-line transaction processing, Internet, intranets, and other storage-intensive applications are causing rapid growth in network storage. Some analysts have estimated that by the end of the century, the number of network connections for server/storage systems will exceed the number of client connections. Put another way, the number of ports within SAN will exceed the number of users (ports) connected to the SAN. Fibre Channel is the leading example of a high-speed, high-performance SAN architecture.

The different combinations in options are distinguished within a nomenclature system for specifying the options. Figure 15–20 shows the nomenclature. Notice that the speed is typically specified in mega*bytes* per second. The total bit rate, including the overhead for 8B/10B encoding, is shown in parentheses. Not all combinations are possible. Copper cables are typically used in inside a wiring closet to connect equipment over a short distance, between cabinets over distances of less than 25 meters.

Fiber is used for longer distances and higher data rates. Unlike most network applications, the preferred multimode fiber is 50/125-μm. The 62.5/125-μm fiber is recommended for data rates up to 531 Mbps.

FIGURE 15–20
Fibre Channel
nomenclature

Example:
100-SM-LC-L

— Long Distance
— Low-cost, long wavelength laser
— Single mode
— 100 mbytes

Speed
400 400 Mbyte/s (4250 Mbps)
200 200 Mbyte/s (2125 Mbps)
100 100 Mbyte/s (1063 Mbps)
50 50 Mbyte/s (531 Mbps)
25 25 Mbyte/s (266 Mbps)
12 12.5 Mbyte/s (133 Mbps)

Cable
SM single-mode fiber
M6 62.5-μm multimode fiber
M5 50-μm multimode fiber
TV 75-ohm coaxial video cable
MI miniature coaxial cable
TP 150-ohm shielded twisted-pair cable

Transmitter
LL long-wavelength laser
LC low-cost long-wavelength laser
SL short-wavelength laser (with OFC)
SN short-wavelength laser (without OFC)
LE long-wavelength LED
EL electrical

Distance
L long 2 to 10 km
I intermediate
S short distance

Long wavelength = 130-nm window. Short-wavelength = 760 nm to 850 nm.

Fibre Channel has three main application topologies, as shown in Figure 15-21. These are point-to-point, switched fabric, and arbitrated loop.

The point-to-point link should be obvious. Two devices are connected directly to one another over a bidirectional link. The original concept of Fibre Channel was as a point-to-point link between devices. The main allure was speed, but this simple architecture did not have the rich flexibility required.

The switched fabric uses hubs and switches to interconnect devices in the same manner as a LAN. Where a LAN is typically used to connect PCs, servers, and peripherals, the Fibre Channel fabric typically connects storage devices and the like. However, because the concepts of the Fibre Channel fabric and the Ethernet star-wired network are similar, you could build a LAN based on Fibre Channel.

The fabric is usually pictured as a cloud, since each attached device needs only to manage the point-to-point connection between itself and the fabric. It does not need to

FIGURE 15-21

Fibre Channel
topologies

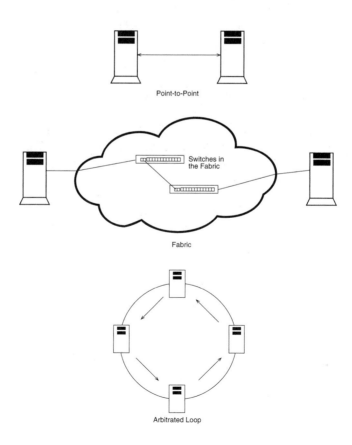

Point-to-Point

Fabric

Arbitrated Loop

be concerned with the specifics of the fabric. The fabric may be a single multiport switch or could be a more involved series of hubs and switches. For example, the fabric could connect different Fibre Channel subnetworks.

The advantage of the fabric is the possibility of flexible arrangements. The drawback is that this is the most expensive approach because of the costs of switches and other fabric devices.

The third Fibre Channel topology is the arbitrated loop, often abbreviated FC-AL. FC-AL is a low-cost approach to connect up to 127 devices without hubs or switches. The devices are connected serially in a loop. Only one set of ports can communicate at a time. Information travels around the loop, with each device checking the address of the frame. If the message is addressed to the device, the device processes it. If not, it passes it on to the next station. The loop, then, is a series of point-to-point links, with one link active at a time.

Fibre Channel topologies can be combined. For example, an arbitrated loop can itself be connected to the fabric through a switch. In addition to providing its own upper layer protocols, Fibre Channel can also carry other protocols, including ATM, SCSI (Small Computer System Interface), HIPPI (High-Performance Parallel Interface), Ethernet, IP (Internet Protocol), and others. SCSI and HIPPI, in particular, are widely used for storage subsystems and other large computer systems.

In terms of fiber optics, Fibre Channel offers various options to allow users to meet various performance and price goals. Figure 15-22 shows the speed and distances possible for different options. Notice that Fibre Channel uses both short- and long-wavelength lasers. The short-wavelength devices include both low-cost CD lasers, VCSEL, and Fabry-Perot lasers operating at 780 to 850 nm. CD lasers use the 780-nm window. The 850-nm window is becoming increasingly popular with the growth of VCSEL. The transmitter/fiber choices work out like this:

Single-mode fiber:	1300-nm long-wavelength laser
50/125-μm fiber:	780 to 850-nm short-wavelength laser
62.5/125-μm fiber:	1300-nm long-wavelength LED

Fibre Channel specifies the SC or Volition connectors for fiber. Plug-in interfaces make it easier to swap in different interfaces to accommodate copper or fiber at different speeds. Fibre Channel uses a media-independent interface. This is a fancy way of saying that the interface between the physical layer and higher layers is independent of the transmission medium. This allows plug-in modules to be designed for different transceivers and media. Called gigabit link modules, these plug-in modules allow either an optical or copper interface, with any of the allowed media, to be used with no redesign of the equipment. It may be slightly cheaper to "hard wire" the interface. In addition, media converters are also available.

While Fibre Channel allows subgigabit speeds from 533 Mbps down to 133 Mbps, the main application of Fibre Channel appears to be for gigabit links. The two great benefits

FIGURE 15-22
Various transmission distances for Fibre Channel over fiber

Fiber	Transmitter	Transmission Distance (max)					
		4.250 Gbps	2.125 Gbps	1.063 Gbps	533 Mbps	266 Mbps	133 Mbps
SM	LW Laser	10,000 m	10,000 m	10,000 m	10,000 m	10,000 m	
50-µm MM	SW Laser			500 m	1000 m	2000 m	
62.5-µm MM	LW LED					1500 m	1,500 m

of Fibre Channel, as with other networks offering fiber, is high bandwidth and longer transmission distances. It appears that users are opting for these benefits. Keep in mind that Fibre Channel can also be used to connect several disk drives within a single cabinet. These interconnections will probably be copper because of the very short distances involved. However, as Fibre Channel migrates toward the higher 2.125- and 4.250-Gbps rates, fiber—particularly parallel array links—will become attractive in inside-the-box uses.

The OFC mentioned in Figure 15-20 means *open fiber control*. The short-wavelength lasers used in Fibre Channel are often not Class 1 eye-safe devices. Therefore, they present a hazard if a person looks into a transmitter port or an energized fiber. Normally, the system is closed, with the transmitter in one device connected to the receiver in another device. But since Fibre Channel is not necessarily applied in a protected environment, users should be protected.

OFC is a method of shutting off the laser if the circuit is opened. Consider that a Fibre Channel link has two conditions: transmitting data or not transmitting data. Even when the link is not transmitting data, it continuously transmits pulses to the other device. As long as the device is receiving light (either data or the OFC pulses), it knows the connection is closed. If the receiver does not receive pulses, it immediately transmits an alarm back to the sending device and the sending device shuts down. OFC, then, is a safety interlock that prevents the laser from transmitting over an open link. The response time to turn the laser off is on the order of 2 to 4 milliseconds (depending on the link speed). A faster link requires a faster turnoff time.

ATM

Asynchronous transfer mode (ATM) was thought by many to be the network that would emerge the winner in the high-speed wars. ATM offered a number of features that seemed to make it the answer for both LAN and WAN applications. It offered not only seamless integration of LANs and WANs; it also promised a robust QoS for delivery of data, voice, and video; it could work with existing LANs such as Ethernet; and it was scaleable. ATM did not prove the be-all end-all network its originators had hoped. Gigabit Ethernet, for example, offered an upgrade path to higher speeds for Ethernet that made network administrators more comfortable. Nevertheless, ATM is an important technology. If it does not replace Ethernet and FDDI as in the LAN, it is popular as an edge device connecting LANs to WANs. In addition, it is widely used by telecommunication companies.

ATM's basic operating speed is 155 Mbps. It can also work at 100 Mbps, using the same chipsets signaling techniques as FDDI, and scale up to 622 Mbps and gigabit speeds. Runs to the desktop can use 51 Mbps.

Equally important, ATM scales upward to higher speeds and is compatible with SONET. SONET is the fiber-optic scheme used by telephone companies for high-speed long-distance communications (and will be discussed in the next chapter). SONET speeds start at 51 Mbps and scale upward in multiples of the basic speed to 10 Gbps and above. SONET does not have a top-end defined, but today's SONET systems top out around 10 Gbps. The attractiveness of this capability is that both LAN users and the telephone companies can adopt ATM. ATM uses SONET as the physical layer. Because wide-area networking is becoming evermore important, ATM provides a straightforward way to use the telephone system as part of your network. Standards are being set to allow ATM to operate at 622 Mbps and 2.48 Gbps (both of which are SONET speeds). Such speeds, however, will require some reconsideration of building cabling. At such speeds, fiber will be the preferred medium, with single-mode fiber required for long backbone interconnections.

Second, ATM is a circuit switching network. ATM can establish a virtual circuit between two stations, allocating and guaranteeing the bandwidth required for the applications. This virtual circuit is very similar to a telephone call. The telephone company switches your call to the other end. As far as you and the other party are concerned, you have a dedicated circuit for as long as the phone call lasts. When you hang up, the circuit is broken. The difference between the voice call and the ATM virtual circuit is that ATM can guarantee exactly the amount of bandwidth required. As many users require varying amounts of bandwidth, ATM hubs must allocate the bandwidth dynamically. It can distinguish between time-critical and other requirements to prioritize communications between stations.

Third, ATM offers compatibility with existing Ethernet networks through emulation techniques. It is possible for existing low-speed networks and high-speed ATM to coexist. Like FDDI, ATM can serve as a high-speed backbone connecting other types of LANs.

Ethernet and FDDI use variable packet sizes. An Ethernet frame, for example, can range in length from 64 to 1518 bytes, while an FDDI frame can range from 20 to 4506 bytes long. One drawback to variable-length frames is that it becomes difficult to determine how long a frame takes to transmit on the network. In contrast, ATM uses a fixed length 53-byte cell. The cell is divided into two sections called the header and the payload. The 5-byte header carries addressing information while the 48-byte payload carries the information—voice, data, or video.

ATM defines two types of interfaces: the user interface and the public interface. You can think of the user interface as the LAN-side, supporting both multimode and single-mode fiber. The public side is the interface to the public telephone network and is defined as a single-mode interface. If ATM is used only as a LAN, all interfaces are user interfaces. If ATM is used only within the telephone system, all interfaces are public. When ATM is used as an edge device to connect the LAN to the WAN, the device will have both user and public interfaces.

From the perspective of transmission media, ATM offers a wide range of options. Besides the standard copper cables and single-mode and multimode fibers, ATM is notable for supporting plastic fiber. Still, the most popular cable choices are shielded twisted-pairs, single-mode fiber, and glass multimode fiber. Figure 15-23 shows examples of fiber options for ATM. For LAN applications, ATM typically runs at 622 Mbps or 1.2 Gbps. Since ATM maps nicely onto SONET, it can operate at SONET rates in long-distance applications and ATM switches are often available with SONET interfaces. Higher speeds are in the process of being defined. For the time being, a public SONET interface can be used to support higher speeds in edge devices.

SUMMARY

- Many fiber-optic applications have been standardized.
- Local area networks use fiber optics.
- Copper and fiber will coexist in networks.
- Traditional networks used shared media; newer versions use switched media.
- Ethernet is a CSMA/CD network operating at 10, 100, and 1000 Mbps.
- Token Ring is a token-passing network running at 4, 16, and 100 Mbps.
- FDDI is a token-passing network operating at 100 Mbps and was originally designed for fiber.

FIGURE 15-23 Some fiber options for ATM transmission

ATM INTERFACE	DATA RATE	FIBER TYPE	DISTANCE
OC-3	155 Mbps	Multimode	2 km
		Single-mode	15 km (intermediate reach)
		Single-mode	40 km (long reach)
		POF	50 m
OC-12	622 Mbps	Multimode	300 m
		Single-mode	15 km (intermediate reach)
OC-48	2.4 Gbps	Single-mode	40 km (long reach)

- Premises cabling is an application-independent approach to wiring a building generically to handle many applications.
- The three main premises-cabling architectures are distributed, centralized, and zoned.
- Most premises-cabling systems are divided into backbone and horizontal cabling.
- The standard premises application limits horizontal runs to 90 meters. However, centralized fiber systems can run 300 meters.
- Structure cabling systems are used both for premises cabling and data center cabling.
- Centralized cabling architectures best exploit the advantages of fiber.
- ESCON is a fiber-optic channel for attaching equipment to a mainframe computer.
- Fibre Channel is an approach for connecting computer equipment at data rates up to 4.25 Gbps.
- Fibre Channel has three flavors: point-to-point, fabric, and arbitrated loop.
- Fibre Channel can be used for storage-area networks.
- ATM is a robust network that can be used for both LAN and WAN applications.

 Review Questions

1. Define the operation of CSMA/CD in controlling access to a network.
2. Describe the difference between a shared-media network and a switched-media network. Explain whether Token Ring, Ethernet, and ATM uses shared or switched media.
3. What are the two types of cable used in a premises-cabling system.
4. What is the maximum distance recommended for horizontal cabling in a distributed premises-cabling network? What is the difference in recommended distance for fiber versus copper in a distributed system?
5. How does a centralized network take better advantage of the characteristics of optical fibers?

6. While Fibre Channel was designed for connecting data networking equipment, it can also be used as a LAN. Which Fibre Channel topology is most suited to a LAN?
7. ESCON systems are used to connect what kinds of equipment?
8. What are the two main data rates of ATM?
9. What is the main difference between an ATM cell and an Ethernet frame?
10. What is the purpose of quality of service?

chapter sixteen

FIBER-OPTIC SYSTEMS

telecommunications and broadband applications

This chapter looks at two important applications of fiber optics: telecommunications and residential broadband. In some respects the two overlap but for convenience, we will divide them up and treat them separately. For telecommunications, we will focus primarily on long-distance communications. For broadband, we will discuss the delivery of services to the home. This typically means cable television (CATV), but increasingly it includes high-speed Internet access and other "value-added" services beyond telephones and basic cable TV.

TELECOMMUNICATIONS AND SONET

Outside of the local loop that delivers telephone services to homes, fiber rules. Nearly all transmission between telephone central offices is done on optical fibers. While some copper plant may remain, fiber has been the preferred cable choice since the 1980s because of its low loss and high bandwidth. SONET is the standard for high-capacity optical telecommunications. As discussed in Chapter 3, SONET is based on multiples of a base transmission rate of 51.84 Mbps. Thus, OC-3 has a transmission speed of 155.52 Mbps, OC-48 operates at 2.48 Gbps, and OC-192 runs at 9.95 Gbps. SONET is deployed mainly in North America. The international version of SONET is called SDH, for Synchronous Digital Hierarchy. The equivalent of the OC is the STM or Synchronous Transport Module. OC and STM differ from one another in one respect: their numbering systems. The OC number is always three times that of the SDH number: OC-12 = STM-4, for example.

Figure 16-1 summarizes significant SONET/SDH data rates.

The trend is to increase SONET data rates by a factor of four. The rates jump from the OC-3/STM-1 rate of 155 Mbps to the OC-12/STM-4 rate of 622 Mbps or from the OC-48/STM-16 rate of 2.5 Gbps to the OC-192/STM-64 rate of 10 Gbps. The DS0 column, added for perspective, shows the number of basic 64-kbps voice channels that can be carried at each level. Thus, OC-192/STM-64 can transport nearly 130,000 calls over a single fiber operating at a single wavelength. As of early 1998, the fastest commercially available systems offered OC-192 rates of 10 Gbps. The fourfold jump to 40 Gbps presents significant technological challenges in both the optical and electronic parts of the system. Still, companies are working on making 40 Gbps a reality, and even higher speeds may occur in the future. Fortunately, as you will see shortly, there are optical methods for increasing capacity in a system without increasing the SONET speeds.

SONET is at the lowest layer of the OSI hierarchy, the physical layer providing the transport medium for data. SONET transmits data in frames. The organizational structure is called Synchronous Transport Signal Level 1 (STS-1) and is shown in Figure 16-2. An STS-1 frame overlays data on an OC-1 optical signal operating at 51.84 Mbps. The signal is typically shown as a two-dimensional array of 90 columns by 9 rows. Each cell in the array represents one byte, so there are 810 bytes or 6480 bits in each frame. The first three columns are used for transport overhead—that is, they are used for network control and monitoring. The remaining 87 contain the payload, the data being transported. Frames have a repetition rate of 8000 frames per second and a duration of 125 μs. If you multiply the 6480 bits per frame times the 8000 frames per second, you get the OC-1 data rate:

6480 bps × 8000 fps = 51,840,000 bps

Even though the STS frame is shown as an array, it is transmitted serially. Row 1 is transmitted first, then row 2, row 3, and so on through row 9.

FIGURE 16-1
SONET and
SDH rates.

LEVEL		TRANSMISSION	DS0 CHANNELS
OC (SONET)	STM (SDH)	RATE	(VOICE CHANNELS)
OC-1	—	51.84 Mbps	672
OC-3	STM-1	155.52 Mbps	2016
OC-12	STM-4	622.08 Mbps	8064
OC-48	STM-16	2.488 Gbps	32,256
OC-192	STM-64	9.953 Gbps	129,024
OC-768	STM-256	39.813 Gbps	516,096

FIGURE 16-2
STS-1 frame

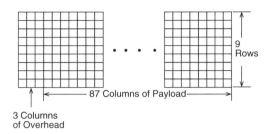

87 Columns of Payload

9 Rows

3 Columns of Overhead

Groups of STS-1 frames are mixed into higher levels byte by byte. The incoming streams of parallel STS-1 data must be synchronized with one another and have the same structure. The multiplier combines them one by one into a serial data stream.

SONET carries many types of traffic. One type that is of interest is ATM, which is designed to work closely with SONET. ATM runs on top of SONET.

Figure 16-3 shows the application of SONET in point-to-point and ring topologies. The earliest application of SONET was as a point-to-point link. Today, SONET is typically a ring. A point-to-point link is more prone to failure than a ring. A cable break between two points means an involved rerouting of signals. A ring, on the other hand, offers more than one path.

SONET uses one of two types of rings: line switched and path switched. A line-switched ring switches the entire optical line. A path-switched ring switches individual paths.

Path switching sends traffic both ways around a two-fiber ring for redundancy. The receiving end monitors both signals and uses the better one. One fiber serves as the working fiber and the other as a protection or backup fiber. Because two fibers are always in use, a path-switched system has less capacity than a line-switched system.

FIGURE 16-3
SONET in
point-to-point
and ring
applications

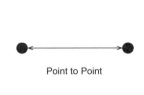

Point to Point

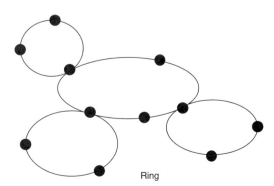

Ring

FIGURE 16-4
SONET ranges

RANGE	DISTANCE	LOSS (dB)	SOURCE
Short Reach	up to 2 km	7	LED or laser
Intermediate Reach	up to 15 km	12	Low-power (50 mW) laser
Long Reach	up to 40 km	10–28	High-power (500 mW) laser

Optically, SONET uses single-mode fiber and defines three different ranges, with differing requirements for light source, loss budget, and transmission distances as shown in Figure 16-4.

TOWARDS ALL-OPTICAL SYSTEM

Fiber-optic technology offers great increases in bandwidth and transmission distances. System designers in the 1980s thought that the single-mode fibers being installed offered an inexhaustible capacity waiting to be exploited. But simply increasing the speed at which a fiber-optic transmission system sends a single channel is becoming difficult. Limitations in fibers and sources limits the speed and distances. Data rates for a single channel include 2.5 Gbps (OC-24) and 10 Gbps (OC-96). Higher data rates present challenges in obtaining fast, narrow wavelength lasers and in exploiting the embedded cable plant. As you saw in Chapter 6, zero-dispersion fibers have limitations that increase noise levels. While nonzero-dispersion-shifted fibers are preferred, the reality is that a considerable amount of zero-dispersion-shifted fiber is installed today.

Four advances in technology are changing the way fiber-optic communications systems are being deployed:

- Dense wavelength-division multiplexers
- Optical amplifiers
- Optical add/drop multiplexers
- Optical cross connects

These devices are important for two reasons. First, they enable high-speed, long-distance transmissions without repeaters. Second, they are important steps toward all-optical transmission systems. By eliminating the need to transform optical signals to electrical signals at various points in the transmission path, costs can be lowered and more efficient systems designed and built. We will discuss each of these devices in turn and then show how they all fit together.

DENSE WAVELENGTH-DIVISION MULTIPLEXERS

SONET is a time-division multiplexing (TDM) scheme. TDM interleaves bytes of data from different signals and sends them serially over the fiber. Thus if you have three data streams A, B, and C, they will be sent over the optical fiber as A1, B1, C1, A2, B2, C2, A3, B3, C3, and so on.

In a dense WDM system, several different optical channels are transmitted over an optical fiber. At the receiving end, the DWDM demultiplexer separates the channels. Figure 16-5 shows an example of a DWDM system. Each wavelength at the input represents an OC-48 channel operating at 2.5 Gbps as four separate 2.5-Gbps signals. The multiplexer sends all four channels over a single fiber. Thus, the fiber carries 10 Gbps. At the receiving end, the DWDM demultiplexes the signal back into its four separate wavelengths. The *dense* in DWDM comes from the fact that the entire system operates in the 1550-nm region. Channels are typically separated by 0.8 nm or 1.6 nm. For example, one channel can operate at 1552.527 nm, while another channel operates at 1553.332.

The ITU–T has defined channels covering a wavelength from 1530 nm to 1560 nm. The channels are actually defined in terms of frequency (remember that wavelength and frequency are two ways of describing the same wave). The reference frequency is 193,100 GHz, which corresponds to a wavelength of 1552.527 nm. Each channel is separated by 100 GHz or 0.8 nm. Many DWDM systems use 200-GHz channel separations because the filtering requirements are not as precise. Figure 16-6 shows the channel assignments. The 100-GHz channel spacing is not the end: Systems with spacings of 50 GHz (0.4 nm) and even 25 GHz (0.2 nm) will allow even greater capacity by packing even more wavelengths into efficient passband of erbium-doped fiber amplifiers from 1530 nm to 1560 nm. In addition, some manufacturers are extending the wavelengths used beyond 1560 nm. The number of wavelengths that will be sent through a single fiber will continue to grow.

FIGURE 16-5

Dense wavelength-division multiplexing

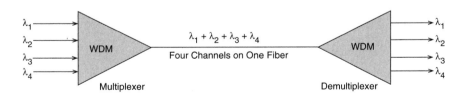

FIGURE 16-6
ITU-T channel spacings for DWDM

WAVELENGTH	FREQUENCY (GHz)	WAVELENGTH	FREQUENCY (GHz)
1532.68	195,600	1547.72	193,700
1533.46	195,500	1548.51	193,600
1534.25	195,400	1549.31	193,500
1535.04	195,300	1550.12	193,400
1535.82	195,200	1550.92	193,300
1536.61	195,100	1551.72	193,200
1537.40	195,000	1552.52	193,100
1538.19	194,900	1553.33	193,000
1538.98	194,800	1554.13	192,900
1539.77	194,700	1554.94	192,800
1540.56	194,600	1555.75	192,700
1541.35	194,500	1556.55	192,600
1542.14	194,400	1557.36	192,500
1542.94	194,300	1558.17	192,400
1543.73	194,200	1558.98	192,300
1544.53	194,100	1559.79	192,200
1545.32	194,000	1560.61	192,100
1546.12	193,900	1561.42	192,000
1546.92	193,800	1562.23	191,900

DWDM enables great increases in capacity without the need for increasing the basic rate through a fiber. If a fiber has a transmission limit of 10 Gbps, you can still increase the speed to 40 Gbps by using a four-channel WDM. Figure 16-7 shows the increases achievable with DWDM.

The wavelengths multiplexed through a DWDM system do not have to operate at the same rate. You can mix different rates to satisfy different needs.

Figure 16-8 shows a DWDM system. The system has a capacity of 3.2 Tbps—that's *tera*bits or trillions of bits. The system handles up to eight fibers, with each fiber having the ability to transmit 400 Gbps. The system can handle both 2.5-Gbps and 10-Gbps channels and can multiplex up to 80 channels over a single fiber.

FIGURE 16-7
Dense wavelength-division multiplexing increases the transmission capacity of an optical fiber

DWDM CHANNELS	CHANNEL RATE	
	OC-48 (2.5 Gbps)	OC-192 (10 Gbps)
4	10 Gbps	40 Gbps
8	20 Gbps	80 Gbps
16	40 Gbps	160 Gbps
32	80 Gbps	320 Gbps
64	160 Gbps	640 Gbps
80	200 Gbps	800 Gbps

OPTICAL ADD/DROP MULTIPLEXER

The DWDM system discussed above and shown in Figure 16-5 is a point-to-point application. All channels are multiplexed onto the fiber at one end and demultiplexed at the other. What if you want to add or remove channels somewhere in between? To do that, you need an optical add/drop multiplexer (OADM), as shown in Figure 16-9. Channels can be added to a fiber or channels can be removed. Some channels, called express channels, are through channels that are not affected.

FIGURE 16-8
A dense wavelength-division multiplexing system (Courtesy of Lucent Technologies)

FIGURE 16-9
Optical add/drop
multiplexing

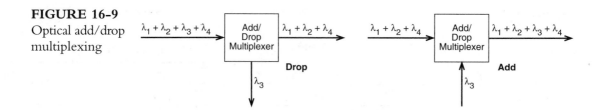

Simple OADMs allow one specific channel to be dropped or added. More complex ones allow the channels to be selected. The ability to select channels provides greater flexibility.

OPTICAL FIBER AMPLIFIERS

OFAs greatly extend the distances that can be achieved in long-haul systems. In addition, they change many of the ways we view optical systems. Since an OFA can increase optical power significantly, overall attenuation becomes less of an issue. Because OFAs can be applied at the transmitter, in midspan, or at the receiver, signal power is more easily handled. As shown in Figure 16-10, the combination of DWDMs and OFAs can dramatically change the physical structure of a fiber-optic system. Electronic regenerators are eliminated, span distances are increased, and possible transmission distances are increased.

OPTICAL CROSS CONNECTS

In a fixed system, one fiber may be connected to another: One input connects to one output. If you have a patch panel, you can manually change the input and output fibers. An optical cross connect allows the input and output fibers to be automatically selected without the need to plug or unplug connections. Figure 16-11 shows the idea of an optical cross connect.

Why optical cross connects? They allow signals to be rerouted around breaks. If your signal is passing through Winston-Salem on the way to Orlando and a break occurs, you can reroute the signal through Charlotte. Thus changeovers already occur with copper-based systems or electronically in a fiber-based system.

An optical cross connect should have the following capabilities:

- All optical: no signal regeneration
- Switch an optical signal (fiber-to-fiber switching)
- Switch individual wavelengths
- Perform add/drop of individual wavelengths

FIGURE 16-10
The combination of DWDMs and OFAs increases the flexibility and capacity of optical systems

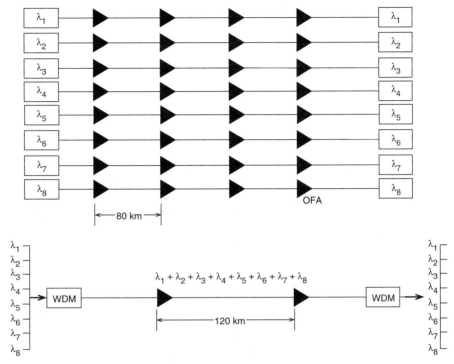

- Perform wavelength conversions (sometimes called wavelength interchange or wavelength translation)
- Be scalable to handle differing numbers of inputs and outputs

We'll briefly discuss each issue.

All optical, no regeneration. Optical cross connects don't yet have all these capabilities in a single unit. For example, signal regeneration is often performed at the cross connect. The reason is simple: Each optical amplifier along the transmission path adds some noise to the signal. The additive noise limits the number of hops or optical amplifiers that a signal can pass through before it needs to be regenerated. In early 1998, the number of hops was typically 5. Improvements in optical amplifiers, lasers, and other components will likely increase the number of hops in the future.

The cross connect makes a convenient place to regenerate the signal. You don't necessarily know where the signal is coming from or where it is going after it leaves, so you don't know how many optical amplifiers it has passed through. Regenerating the signal at a central location as a cross connect simplifies system planning.

FIGURE 16-11
The optical cross connect allows a signal on a input fiber to be switched to any output fiber

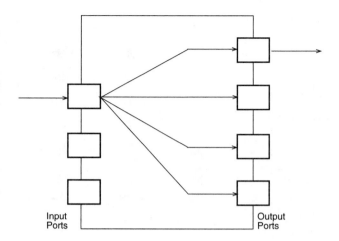

Input Ports

Output Ports

Switch an optical signal (fiber-to-fiber). Switching the entire DWDM signal is the simplest form of a cross connect. It is of use mainly for restoration work, to reroute the signals around a break in the line.

Switch individual wavelengths. The ability to switch individual wavelengths allows much more sophisticated control and management of optical signals. Any wavelength at any input port can be switched to any output port.

Add/drop. Once you have the ability to distinguish and control individual wavelengths, it becomes easier to perform add/drop multiplexing.

Wavelength conversion. When dealing with individual wavelengths, the possibility arises that the same wavelength must be switched onto a line. In this case, you must convert one of the wavelength to another wavelength. Suppose, for example, you want to drop off the "red" wavelength coming into the cross connect on port 1. Then, on port 3, you also need to drop the "red" wavelength. One of these wavelengths will have to be converted to another color so that both can be multiplexed onto the same output fiber. The conversion can be done optoelectronically with lasers or optically through a number of techniques. The four-wave mixing, described in Chapter 6, for example, results in new wavelengths.

Scalability. How many inputs and outputs do you need? How many multiplexed wavelengths will be handled. Technology for a four-wavelength, 2×2 cross connect may not be suitable for a 16-wavelength 16×16 cross connect. A cross connect should be scalable so that the user can expand capabilities as the need arises.

One other note about cross connects: Switching does not have to be lightning fast. A cross connect can take several milliseconds to switch a signal from one fiber to another. Nor is switching a rapidly reoccurring phenomena, where signals continually switch between ports. Don't confuse a cross connect with a network switch, such as an ATM or

Ethernet switch. These must switch signals between different ports steadily at very high data rates.

BROADBAND

By broadband we mean methods of delivering services over a channel broad enough to support multiple services. The channel may also be viewed as a number of separate channels: CATV, for example, can deliver 80 channels of television programming to your home. Within the catchbag of broadband, we can also obtain the services listed in Figure 16-12. The main thing is that these require more bandwidth than typical telephone service.

First, let's get two terms straight: service provider and subscriber. The cable TV and telephone company are the service providers. The family is the subscriber since they pay for the services provided by the service provider. In many respects, the cable TV companies and telephone companies are competing today to deliver these broadband value-added services. As you might expect, the CATV providers have an advantage in providing television services. But the Internet can become an important vehicle for delivery of services and telephone companies have the advantage here. In addition, high-speed telephone services will allow telephone companies to deliver video and similar services. Both types of companies want to deliver some or all of these services and obtain the revenues.

FIBER IN THE LOOP

Service providers are concerned with "the last mile" from the service provider backbone to the home. Backbone technology is well understood, both technically and economically. But how to cost-effectively bring new services to the home is still a matter of debate. Service to homes is called the *local loop,* as opposed to interoffice links between telephone central offices or between CATV headends. So we speak of fiber in the loop (FITL). There are many approaches to how to deploy fiber and how far into the loop to bring it. This has resulted in a variety of terms and acronyms for the different approaches:

FITL: The general term.
FTTC: Fiber to the curb. Fiber is run to a small neighborhood of about a
         dozen homes or less. Coaxial cable runs to the home.
FTTH: Fiber to the home. Fiber is run to a network interface unit (NIU) on
         the outside of the house.
FTTB: Fiber to the basement. This is used for apartment buildings and high
         rises. The fiber is brought to the basement of the building and coaxial
         cable is used to distribute services throughout the building.

FIGURE 16–12
Typical services
from residential
broadband

Cable TV	80 channels and nothing to watch.
Video on demand (VOD)	Movies or other video when you want them. You order the movie you want to see from hundred or thousands immediately available. Watch the *Blues Brothers* every night or every episode of *Gilligan's Island.* A related, but less expansive version, is near video on demand (NVOD), which allows you to pick from dozens of movies that begin at regular intervals.
Internet access	Surf the Internet at high speeds. By this we mean at speeds faster than what is possible with a voice modem. Speeds can range from 64 kbps to several megabits per second. By contrast, voice modems operate at 28.8, 33.6, or 56 kbps. Most of the following services can be obtained over the Internet, although the service provider can provide them by other means.
Information	Get stock quotes, sports scores, weather, or nearly any kind of regular information.
Home shopping	Order pizza from the corner pizzeria or a new computer from an international company. Such buying with on-line electronic commerce is coming into its own in the late 1990s.
Video conferencing	Talk to and see others through a video camera connected to a computer or other device.
Distance learning	Get an education by connecting with a remote location, say a college across the state or across the globe. The ability to have live, interactive video is helpful to obtain maximum benefit.
Work at home	So-called telecommuting is a growing trend. The ability to have high-speed connection for voice, video, and data to the office and customers makes working at home while being electroni-

There is no question that it is desirable to bring fiber optics directly to and even into the home. The problem is cost. It is simply too expensive to rewire the telephone or cable TV grid to bring fiber to the home. Part of the problem is technical: Fiber components are still relatively expensive. However, prices are dropping as technical innovations lower components costs and allow the cost of installation to fall. In addition, advances like fiber amplifiers make it easier and less expensive to deliver sufficient optical power to every home. But replacing the existing copper infrastructure with an optical one is expensively daunting. Service providers (the telephone company or CATV company) simply cannot recover the cost within a price structure consumers are willing to pay.

As technology improves and prices decrease, fiber may first be found in new construction. If the cost is low enough, service has to be provided one way or another. It's not a question of ripping out one set of cable and installing a new one. If you are running brand new cable, eventually fiber will become economically attractive.

Today, however, the most common method is called hybrid fiber-coax (HFC). In the HFC system, fiber is used in the backbone, while coaxial cable is used for local distribution to the home. There are different approaches to HFC that primarily involve how close to the home the fiber goes and how many homes are served by each node. A typical system operates over a range from 55 MHz to 750 MHz. Standard analog television is delivered from 55 MHz to 550 MHz. The 200 MHz from 550 to 750 MHz is used for digital delivery of television and other needs. With compression, this band can handle over 300 channels.

By now, you should recognize the advantages of fiber in the backbone. The low loss and high bandwidth allow long unrepeatered transmission. This dispenses with expensive coaxial amplifiers every 1000 to 2000 feet.

Figure 16-13 shows a typical HFC system. The point of the signal origin in the distribution system is called the headend. This can be a cable TV transmission facility (or a telephone company switching office). The fiber carries the system to a point called a node, where coaxial cable picks up. As you can see, there still can be coaxial amplifiers in the path, but considerably fewer than in an all-copper system. Again, the ideal situation is to reduce two things: the number of subscribers served by each node and the number of amplifiers and other electronics in the path between headend and subscriber. Perfection, then, is reached when each subscriber is served by fiber.

FIGURE 16-13
Hybrid fiber-coax system

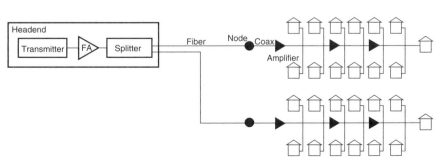

Each node serves between 500 and 2500 homes.

COPPER, FIBER, WIRELESS

While fiber is certainly attractive, it is not the only answer. Both wireless systems and copper system can deliver high-speed broadband services to one degree or another.

The obvious example is direct satellite which can deliver hundreds of television channels. Satellite can also be used to access the Internet. At the same time, digital cellular systems can deliver a host of services beyond telephony: e-mail, paging, and so forth. Wireless allows service providers to skirt the problems of upgrading the infrastructure, installing new cable, or renting existing cable from telephone companies.

The ordinary twisted-pair copper cables has sufficient bandwidth to carry several megabits per second. Digital subscriber line (DSL) technology brings digital capabilities to the local loop.

There are several varieties of DSL. For most consumers, the most important is asymmetric DSL or ADSL. ADSL allows data rates, from the service provider to your home, of anywhere from 1 to 7 Mbps. The rate depends both on how the service provider implements the technology and on the distance from the home to the telephone switching office. A typical maximum distance is 18,000 feet. From the home to the telephone office is much slower: about 640 kbps. The maximum speed differs with direction and thus the technology is called asymmetrical. But consider the needs of most applications. You need fast speeds coming to you in order to view a movie or enjoy the latest and greatest the Internet has to offer—the applications you enjoy are bandwidth intensive. From your home, the needs are much more modest: ordering the video you want to see, issuing an address of an Internet site, and so forth. You will be receiving millions of bytes of information, but only sending out thousands of bytes of information.

DSL technologies are attractive because they use the existing copper plant. Telephone companies do not have to rewire. All that is required is a digital modem in your home and a DSL connection at the telephone switching office.

FIBER TO THE CURB

Typically, we think of CATV and telephone service as being two distinct services, delivered to the home over a different cabling plant. Other than historical development—telephone on twisted-pair cable and CATV on coaxial cable—there doesn't have to be a separation. By using fiber to the home, both services are easily accommodated. Figure 16-14 shows an example of fiber to the curb. Broadband services, such as cable television, operate in the 1550-nm band, while narrowband telephone services operate at 1310 nm. The two are kept separate and easily distinguished from one another. This allows the high costs of installing fiber to be lowered in one of two ways (1) the different service providers can share the costs, or (2) the company that installs it can rent access to the other.

FIGURE 16-14
Fiber to the curb

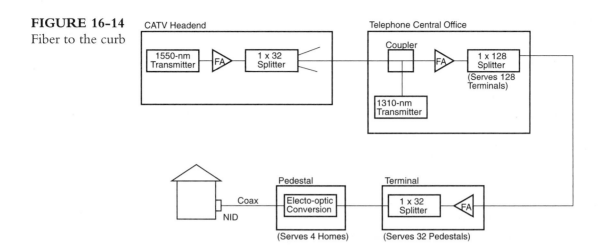

In Figure 16-14, the signal at the CATV headend is first amplified by a high-power fiber amplifier and then split into 32 fibers. For simplicity, the figure only shows the path through one fiber. Remember that at each splitter, the signal is divided among several fibers. The 1550-nm signal from the headend travels to the telephone central office, where a wavelength-division multiplexer couples the signal with a 1310-nm telephone signal. The 1310/1550 signal is again amplified, split, and transmitted to a terminal in the service area. The terminal, which can serve 128 homes, again amplifies the signal before splitting it. The signal travels to a pedestal, where it is split into four channels, each going to a home.

You can see that there is considerable amplification and splitting of the signals along the path. The capacity of the system is 524,288 homes from a single transmitter in the CATV headend and 16,384 homes from the telephone central office. But the signals do not have to be electronically regenerated between the source and destination. In this sense, the optical network is passive. There is electrical power required to operate the optical amplifiers and other system components, but the signal itself remains optical from the headend to the final pedestal.

Not all fiber systems use optical amplifiers. The alternative is to use multiple transmitters. For example, in the figure, a single laser and optical amplifier are used to drive 32 fibers. Without the amplifier, splitting the signal among so many fibers would be impractical. Instead, many lasers would be used. Depending on the system and its loss budget, each laser could service one or more fibers. Other systems will use combinations of fiber amplifiers and multiple transmitters. It depends on the size of the system, the age of the system, and the various cost tradeoffs between amplifiers and transmitters.

At the home (if fiber is brought to the home), typically mounted on an outside wall, is an optical network unit (ONU). The ONU contains transceiver electronics to convert the incoming optical signals to electrical signals and outgoing electrical signals to optical signals. Most recently built homes have a copper version of an ONU called the network interface unit (NIU) or network interface device (NID). NIUs are installed by the telephone companies and serve as a legal and practical demarcation point between the outside plant and the customer premises. In other words, the NIU marks the point that separates problems belonging to the telephone company (the outside plant) from the problems that are yours (the customer premises). A typical NIU contains an interface between the outside wiring and the inside wiring, along with lightning protection and some telephone jacks to allow you to test your connection. If you have problems with your telephone, you can plug a phone into the NID to test the connection. If you get a dial tone, the problem is the inside wiring.

The system can carry both analog and digital signals. If the optical receiver in the ONU has a 1-GHz bandwidth, 80 analog TV channels and hundreds of digital channels are possible. If hundreds of channels are not enough, the system holds the possibility of wavelength-division multiplexing several optical wavelengths for CATV.

The system must also offer a return path, so that you can communicate back upstream to the headend. Such communication includes ordering a movie (VOD), surfing the Internet, purchasing goods and services, video conferencing, and so forth. Different service providers can offer different mixes of services. But since the return path is shared with up to 500 of your closest neighbors, how does the system distinguish one subscriber from another? Put another way, how does the system know to deliver *Star Wars* to your house and not *The Best of Barney* that your neighbor ordered. Each subscriber has an address, similar to the way each computer connected to a LAN has an address. The system uses a fast burst mode to quickly deliver your requests. And like Ethernet, the system uses a collision detection scheme to handle multiple subscribers transmitting at once. If a collision is detected, each subscriber's system will wait a short time and try again.

Although there may be many homes trying to share the return path, the transmission needs are not demanding. Ordering a video requires a few hundred bytes. An Internet request for a new page, too, requires only a few hundred bytes. The bandwidth requirements on a broadband system are typically quite asymmetric and it is easy for many subscribers to share the return path. The return path can be over the same downstream fiber or over a separate fiber dedicated to the return signals.

If fiber is brought to each individual house, the NID becomes an ONU. An ONU is a bit more complicated since it contains the electro-optic interface, including a demultiplexer for separating the 1310-nm telephone and 1550-nm CATV channels, a receiver for converting the optical signal to an electrical one, and a transmitter for upstream transmissions. One important aspect of the interface is that it be bidirectional: you must be able to send signals back upstream, either a telephone call, Internet communications, video orders (such as video on demand or pay per view), and so forth.

ANALOG AND DIGITAL BROADBAND

The first generation of HFC systems used 1310-nm technology with a directly modulated laser. The industry has moved to 1550-nm systems indirect modulation (using electroabsorption or lithium niobate modulators) because of the advantages of this operating window. Still, 1310-nm systems are widely used and capable of carrying 110 analog television channels over spans as long as 50 km.

Although most HFC systems are analog, digital transmission is coming on strong. Analog systems have some specific needs that are different from digital systems. A key requirement is high linearity so that the output of the laser is identical to the analog input. Many lasers used for digital applications do not provide the required degree of linearity. Specially designed analog lasers are used.

The special requirements of analog systems often mean an expensive DFB laser is required to achieve the required linearity. For digital systems, a lower-cost Fabry-Perot laser can suffice because the demands of a digital signal are less stringent.

System noise is another area where analog and digital differ significantly. The quality of a digital transmission can be evaluated by the BER: the probability of a single bit being misinterpreted by the receiving electronics. While BER is related to the SNR, BER is usually the key specification. To obtain a suitable BER of 10^{-12} to 10^{-15}, an SNR of only 10 dB or so is required.

Analog systems have different requirements and different sources of noise. Signal quality is given by the carrier-to-noise ratio (CNR), which is the ratio of the carrier signal power to the noise power. The CNR in an analog system must be much higher than the corresponding SNR in a digital system: 50 dB to 10 dB.

Two common types of noise in an analog video system are composite second-order (CSO) and composite triple-beat (CTB) distortion. These can appear as rolling or intermittent diagonal lines on your TV. CSO and CTB are part of the nature of the analog transmission. They are caused by the way the different analog frequencies of the different channels interact with one another. Such interactions can distort the picture. CSO and CTB can be calculated, although for an 80-channel system the calculations can become complex and require a computer. But once you know the worst-case distortion from CSO and CTB, you can design against it. A common solution is predistorting the laser drive signal so that the predistortion and CSO/CBT distortion cancel one another and the effects are kept well within acceptable limits.

In a digital system, you can simply increase the signal power to achieve a higher SNR and better BER. In an analog system, increasing the signal power can actually increase noise. An effect known as stimulated Brillouin scattering (SBS) can cause reflections, increase noise, and reduce the signal strength. Another aspect of noise in an analog system is that once it enters the system, it remains there. A digital system allows noise to be eliminated at the receiver and the pulses to be cleanly recovered. The same cannot be done with noise in an analog system.

Several fiber-related characteristics can add noise or reduce the CNR.

- Attenuation, by lowering the signal strength of the carrier over distance, is an obvious contributor. If noise is added along the path, its power can grow even as the signal power decreases.

- Laser noise is caused by random fluctuations in the intensity of the optical signal. One cause is actual increases and decreases in the output power. A second cause is called *interferometric noise,* which results from multiple reflections of light in the fiber. If the light is scattered back toward the laser and then is re-reflected in a forward direction, it can add to the signal. Laser noise is sometimes called relative intensity noise (RIN), because the noise amplitude is referenced to the average amplitude of the optical signal.

- Dispersion can create considerable distortion in an analog signal if the spectral width is sufficiently wide. Dispersion can be reduced by operating at or near the zero-dispersion point so that the effects of chromatic dispersion are minimized.

- Other sources of noise include the receiver's thermal and shot noise discussed in Chapter 9.

Digital broadband is the future. For example, the FCC has mandated that digital television be in place by 2006. Digital systems can use baseband transmission with time-division multiplexing. For mainly analog-based systems, digital data is modulated onto a carrier. The type of modulation used is quadrature amplitude modulation (QAM). QAM changes both the phase and the amplitude of the carrier into discrete combinations. For example, 16-QAM uses 16 different combinations of phase and amplitude to represent bits. Each specific combination represents four bits, as shown in Figure 16-15. Another scheme, 64-QAM, represents 6 bits for every discrete combination of phase and amplitude. The number of bits that can be represented is equal to 2 raised to the power of the number of bits. For example, 16-QAM is $2^4 = 16$. In other words, if you have 16 phase/amplitude combinations, you can represent 4 bits in each combination. Similarly, 64-QAM is $2^6 = 64$. An important feature of QAM is that it allows you to represent multiple bits for each discrete symbol.

QAM is used for the downstream path to the home. The return path uses quadrature phase-shift keying (QPSK), which is a simpler version of QAM.

For practical application, digital video must be compressed to reduce the required bandwidth. The type of compression used is called MPEG2, after the Motion Picture Experts Group that helped develop the technique. MPEG2 uses sophisticated techniques so that only the changes from frame to frame are delivered. In a typical movie, much of the picture from frame to frame remains the same. You really only need to send the changes, not the whole frame. Doing so considerably reduces the bandwidth requirements. Compression ratios can reach 200:1.

FIGURE 16-15
QAM encoding

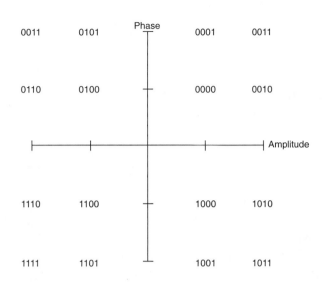

SUMMARY

- SONET is an optical network used for communications.
- SONET uses transmission speeds ranging from 51.84 Mbps to 10 Gbps and beyond.
- The preferred topology for SONET is a point-to-point link.
- DWDM significantly increases the transmission capacity of a fiber by carrying multiple wavelengths.
- The standard channel spacing for optical systems is 100 or 200 GHz, although systems today also operate with 50-GHz spacings.
- DWDM allows terabit systems.
- Add/drop multiplexers allow channels to be added or dropped.
- Optical fiber amplifiers increase possible transmission distances without regenerative repeaters.
- Optical cross connects allow the signal on an input fiber to be switched to any output fiber.
- Fiber in the loop helps enable high-speed broadband services.
- The two main approaches to fiber in broadband networks to the home are hybrid fiber-coax and fiber to the curb.
- Fiber in the loop systems use both analog and digital transmission.

Review Questions

1. What is the base speed of a SONET system?
2. What is the transmission speed of an OC-192 signal?
3. What is the main advantage of dense wavelength-division multiplexing?
4. Assume a DWDM band of 20 nm. How many channels can be wavelength-division multiplexed at a 100-GHz channel spacing? For a 50-GHz channel spacing?
5. List five attractive features of an optical cross connect.
6. Define wavelength conversion. What is its purpose?
7. What is the role of an optical amplifier in a hybrid fiber-coax system?
8. Why do most broadband services to the home not require symmetrical transmission rates? Give an example.
9. Which generally requires a higher optical output in a broadband system, analog or digital systems? Why?
10. What is the advantage of 1550-nm transmission over 1310-nm transmission in a broadband system?

chapter seventeen

INTRODUCTION TO TEST AND OTHER EQUIPMENT

This chapter provides a brief look at some of the equipment commonly used in the field to install, inspect, and maintain fiber-optic systems. It also discusses methods to measure fiber attenuation and connector insertion loss. The equipment discussed is the optical power meter, optical time-domain reflectometer, fusion splicer, polishing machine, inspection microscope, and a few hand tools. Our discussion will be simplified and clarified by looking at specific examples of each.

FIBER-OPTIC TESTING

A great many tests must be performed on optical fibers. A fiber manufacturer must test a fiber to determine the characteristics by which the fiber will be specified. As a quality control measure during manufacture of fibers, the manufacturer must constantly test the fibers to ensure that they meet the specifications. Among such tests are the following:

- Core diameter
- Cladding diameter
- NA *NUMERICAL APERTURE*
- Attenuation
- Refractive index profile
- Tensile strength

Other tests performed on fibers or on fiber-optic cables concern their mechanical and environment characteristics. Mechanical tests such as impact resistance, tensile loading, and crush resistance test the cable's ability to withstand physical and mechanical stresses.

Environmental tests evaluate the changes in attenuation under extremes of temperature, repeated changes (cycling) of temperature, and humidity.

Fiber and cables are specified as the result of such tests. Engineers pick cables suited to their applications on the basis of specifications. Manufacturers of other fiber-optic components—sources, detectors, connectors, and couplers—also test to characterize and specify their offerings.

Other tests are performed during and after installation of a fiber-optic link. These tests ensure that the installed system will meet performance requirements. Splices, for example, must not exceed certain loss values. During installation, each splice must be tested. If the loss is unacceptably high, the fibers must be respliced. This chapter describes how such a test is done with optical time reflectometry.

FIBER TESTING

This section looks at typical testing concerns in a premises-cabling application based on the recommendations found in TIA/EIA-568A for generic cabling systems and TSB-67 on testing. An important concept in testing is distinguishing between the link and channel. The channel is the end-to-end system, including any patch cables at the equipment or work area. The link is the "behind-the-walls" cabling from the equipment-side patch panel and the work-area-side outlet. Figure 17-1 shows the difference between a link and channel.

Cases can be made for testing on either a link level or a channel level. Installers of a building cabling system are usually concerned with the link. The link forms the basic infrastructure for the building cabling. Since the cabling is performance driven, the installer is not concerned with what hubs or what workstations will be used. To a large degree, the cabling in the office or work area is out of the installer's hands. Cabling in the wiring closet or behind the walls is more secure and removed from users—not so the work area cables. The installer, therefore, is concerned with the link, since this is the cabling most easily controlled during and after installation.

Realistically, however, the channel is the real measure of performance. The weakest link in the system are the two patch cables that fall outside the link model. Real-world performance runs from hub to workstation.

Certification is normally done on link level to ensure that the cable is properly installed. Begin with a basic link test. If the link passes, component-level testing is not necessary. If the link fails, test individual components to isolate the fault.

FIGURE 17-1
TIA/EIA-568A
link and channel
models

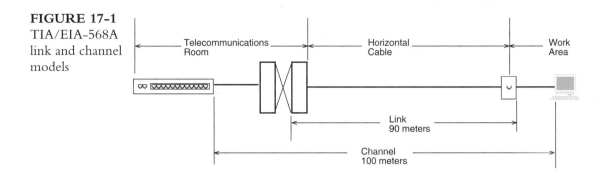

CONTINUITY TESTING

Simple continuity testing can be achieved by a flashlight: Does the light come through the fiber? Fancy flashlights—called visual continuity testers—are available specifically for fiber-optic testing. A red light is easiest to see.

LINK ATTENUATION

Link attenuation is simply the sum of the individual losses from the cable, connectors, and splices. Cable loss can be assumed to be linear with distance. For the 90-meter horizontal cable run, a 62.5/125-μm cable has an attenuation of 0.34 dB at 850 nm and 0.14 dB at 1300 nm. Maximum insertion loss values are 0.75 dB for connectors and 0.3 dB for splices. To estimate the total loss in the link, simply add up the individual losses. Consider a horizontal run with two connectors. The total loss is 1.84 dB:

Cable loss:	3.75 dB/km × 90 m	= 0.34 dB
Connectors:	0.75 dB × 2	= 1.5 dB
Total loss:		= 1.84 dB

TIA/EIA recommends a maximum attenuation for the 90-meter horizontal cable run of 2.0 dB, including the cable and two connectors (one in the telecommunications closet and one in the work area). If passive links are connected serially, the expected loss is the sum of the individual links. Notice that an interconnection includes the connectors on both sides of an adapter. Be careful not to count an interconnection twice. Consider the work area outlet. If you counted this interconnection as part of the horizontal cable, don't count it as part of the work area patch cable.

Notice, too, that TIA/EIA-568A link performance does not include passive elements like couplers, switch, and other optical devices. These can add appreciable loss in the link and must be accounted for if present. Links using such devices will not meet 568 requirements and must be evaluated on an individual basis. For example, 10BASE-FP networks use a passive star coupler to divide light among many fibers. This system would have to evaluated in terms of IEEE 802.3 10BASE-F specifications rather than TIA/EIA-568A specifications.

In estimating cables losses, use the following values:

62.5/125-µm at 850 nm:	3.75 dB
62.5/125-µm at 1300 nm:	1.5 dB
Single-mode outside plant cable at 1310 nm:	0.5 dB
Single-mode outside plant cable at 1550 nm:	0.5 dB
Single-mode inside plant cable at 1310 nm:	1.0 dB
Single-mode inside plant cable at 1550 nm:	1.0 dB

Measured loss must be less than those calculated by worst-case values. However, some care in testing is necessary because the test procedure can significantly affect results.

UNDERSTANDING FIBER OPTIC TESTING

A fiber carries light in modes. The modal conditions of light propagation can vary, depending on many factors. We generally speak of three general conditions: overfilled fiber, underfilled fiber, and equilibrium mode distribution.

- **Overfilled fiber.** When light is first injected into a fiber, it can be carried in the cladding and in high-order modes. Over distance, these modes will lose energy. The cladding, for example, attenuates the energy quickly. High-order modes can be converted into lower order modes. Imperfections in the fiber can change the angle of reflection or the refraction profile of high-order modes.
- **Underfilled fiber.** In some cases, light injected into the fiber fills only the lowest order modes. This most often happens when a low NA laser is used to couple light into a multimode fiber. Over distance, some of this energy shifts into higher modes.
- **Equilibrium mode distribution.** For both overfilled and underfilled fibers, optical energy can shift between modes until a steady state is achieved over distance. This state is termed *equilibrium mode distribution* or *EMD*. Once a fiber has reached EMD, very little additional transference of power between modes occurs.

EMD has several important consequences in light propagation in a fiber and its affects on testing. Both the active light-carrying diameter of the fiber and the NA are reduced. For example, the fiber recommended for premises cabling has a core diameter of 62.5 μm and an NA of 0.275. These are based on physical properties of the fiber. With EMD, we are interested in the *effective* diameter and NA. In a 62.5-μm fiber at EMD, the effective core diameter is 50 μm and the NA is less than 0.275.

So what's the point in being concerned with EMD? The modal conditions of the fiber affect the loss measured at a fiber-to-fiber interconnection. It is easier to couple light from a smaller diameter and NA into a larger diameter and NA. Even so, the light coupling into the second fiber will not be at EMD. It will overfill the fiber to some degree so that EMD will not be achieved until the light has travelled some distance.

In testing, it is important to specify the modal conditions. Testing a connector at EMD can yield significantly different results than testing it under overfilled conditions. A connector can yield an insertion loss of 0.3 dB under EMD conditions and 0.6 dB under fully filled conditions.

In testing, we speak of launch and receive conditions:

- Long launch: Fiber is at EMD at the end of the launch fiber.
- Short launch: Fiber is overfilled at the end of the launch fiber.
- Long receive: Fiber is at EMD at the end of the receive fiber.
- Short receive: Fiber is overfilled at the end of the receive fiber.

EMD conditions can be easily simulated without a long length of fiber. The standard method of achieving EMD in a short length of fiber is by wrapping the fiber five times around a mandrel. This mixes the modes to simulate EMD.

STANDARD TESTS

There are three standard tests applicable to premises cabling.

- OFSTP-14:   link certification with power meters
- FOTP-171:   patch cable certification
- FOTP-61:   link certification with an optical time-domain reflectometer

Let's begin by looking at power meters.

POWER METERS

An optical power meter (Figure 17-2) does just that: measures optical power. Light is injected into one end of a fiber and measured at the other end. Some meters incorporate interchangeable light sources into the meter so that it serves as either a light source or a meter (or both simultaneously). Most meters can display power levels in user selectable units, most often dB or dBm. The dB reading is most appropriate for certification. Because

FIGURE 17-2
Optical power
meter (Courtesy
of Siecor
Corporation)

fiber performance depends significantly on the wavelengths propagated, the light source is important. TIA/EIA-568A makes the following recommendations for wavelengths and testing:

- Horizontal multimode cable: either 850 or 1300 nm
- Backbone multimode cable: both 850 and 1300 nm
- Backbone single-mode cable: both 1310 and 1550 nm

A power meter typically requires two readings. One is used for calibration, to zero the meter, and the other is the actual power reading. Of course, calibration isn't required for each reading; once the unit is calibrated, it can be used for numerous tests. Check the meter's user's guide for guidance on how often to zero the unit.

An optical tester similar to a power meter is the optical loss test set (OLTS). The OLTS is distinguished from the power meter by including the signal injector in the same unit as the power meter.

LINK CERTIFICATION WITH A POWER METER (OFSTP-14)

OFSTP-14 offers two measurement procedures, Method A and B. Both methods are similar.

1. A calibration reading is taken using one or two test jumpers.
2. The jumpers are then connected to the link.
3. The link loss is displayed.

Method A uses the two test jumpers for the calibration, while Method B uses only one. Both methods use two jumpers for the link evaluation. Why the difference? The single-jumper approach of Method B eliminates the influence of the connectors used to interconnect the two cables during the zeroing procedure. The connectors in the middle of the two jumpers essentially mirror the presence of the connectors in the actual link. Test Method A cancels the influence of these connectors during the test. Method B includes these connectors in the test.

Of the two methods, TIA/EIA-568A recommends Method B. Figure 17-3 shows the general setup and sequence for the test. Keep in mind that the figure is general; it does not show patch panels and other intervening couplings that could be included in the test.

TESTING PATCH CABLES (FOTP 171)

This procedure includes four methods, with three optional procedures for each method. We will restrict our discussion to Method B, which is the most suited to premises cabling and is the method typically used by cable assemblers. Figure 17-4 shows the procedure.

This procedure uses the substitution method to evaluate attenuation in a length of fiber. Power through two jumper cables are measured to zero the meter. The cable assembly to be tested is then inserted between the two jumper fibers. The new reading is the fiber attenuation. Notice that the test requires a mode filter on the launch end to simulate EMD in the short length of jumper cable. This can be achieved by wrapping the cable five turns around a mandrel to mix the modes.

FIGURE 17-3
Link certification with a power meter

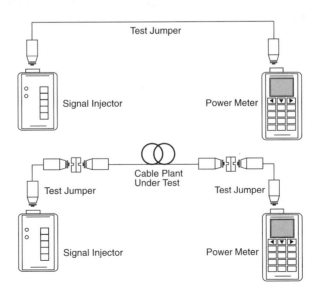

FIGURE 17-4
Testing path
cables

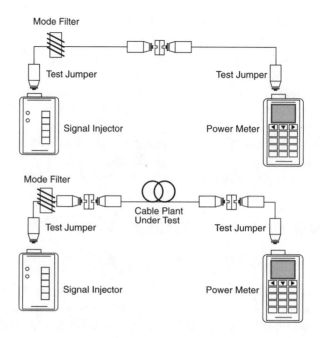

TIME AND FREQUENCY DOMAINS

A signal can be described in terms of the time domain or the frequency domain. This book has used both domains without bothering to distinguish between them. We described the limitations on a fiber's information-carrying capacity in terms of both bandwidth and dispersion. Bandwidth is in the frequency domain; rise-time dispersion is in the time domain. Analog engineers, dealing with analog signals in the frequency domain, talk in terms of frequencies. Digital engineers, dealing with pulses, use the time domain and talk in terms of rise times and pulse widths. We have given a simple method of relating the frequency and time domain in the equations relating bandwidth and rise time.

$$t_r \quad = \quad \frac{0.35}{BW}$$

$$B\sqrt{W} = \quad \frac{0.35}{t_r}$$

OPTICAL TIME-DOMAIN REFLECTOMETER

As its name implies, the *optical time-domain reflectometer* allows evaluation of an optical fiber in the time domain. Figure 17-5 shows an optical time-domain reflectometer.

Optical time-domain reflectometry (OTDR) relies on the backscattering of light that occurs in an optical fiber. Some of the light entering the fiber will be reflected back to the light. *Backscattered* light is that light that reaches the input end of a fiber. Backscattering results from Rayleigh scattering and Fresnel reflections. Rayleigh scattering, remember, is scattering caused by the refractive displacement due to density and compositional variations in the fiber. In a quality fiber, the scattered light can be assumed to be evenly distributed with length. Fresnel reflections occur because of changes in refractive index at connections, splices, and fiber ends. A portion of the Rayleigh scattered light and Fresnel reflected light reaches the input end as backscattered light.

Figure 17-5 shows a simple block diagram of an OTDR unit. Its main parts are a light source, a beamsplitter, a photodetector, and an oscilloscope. A short, high-powered pulse is injected through the beamsplitter into the fiber. This light is then backscattered as it travels through the fiber. The beamsplitter directs the backscattered light to the photodetector. The amplified output of the detector serves as the vertical input to the oscilloscope. Because the power to the detector is extremely small, repeated measurements are made by the electronics of the ODTR; the SNR is improved by averaging the readings, after which the results are displayed.

The OTDR screen displays time horizontally and power vertically. Fiber attenuation appears as a line decreasing from left (the input end of the fiber) to right (the output end). Both the input and the backscattered light attenuate over distance, so the detected signal becomes smaller over time. A connector, splice, fiber end, or abnormality in the fiber appears as an increase in power on the screen, since backscattering from Fresnel reflections will be greater than backscattering from Rayleigh scattering. The quality of a splice can be evaluated by the amount of backscattering: Greater backscattering means a higher-loss

FIGURE 17–5
OTDR block
diagram
(Courtesy
of Photodyne)

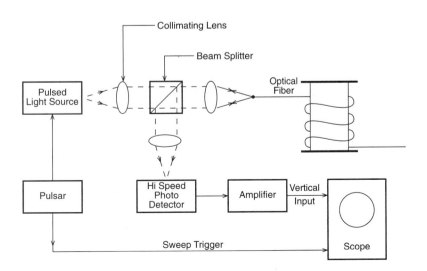

FIGURE 17-6
Optical time-domain reflectometer (Courtesy of Siecor Corporation)

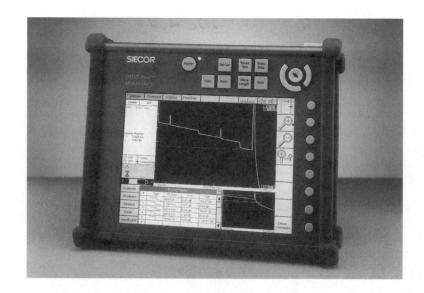

splice. A connector shows both a power increase from reflection and a power drop from loss. The degree of loss indicates the quality of the connection. Figure 17-6 shows a typical display. Notice that it is similar to the power budget diagram in Chapter 13.

Light travels through a fiber at a speed of about 5 ns/m, depending on the refractive index of the core. Time can be correlated to distance by

$$D = \frac{ct}{2n}$$

where D is distance along the fiber, c is the speed of light, t is the round-trip travel time of the input pulse, and n is the core's average refractive index. Most OTDRs use a cursor to mark a horizontal point on the trace and display the distance in terms of both time and physical distance. One can, for example, determine the distance to a splice with a great degree of accuracy, typically to within a foot. The OTDR measures the distance along the fiber, not necessarily the length of the cable, however. If the fiber is stranded around a central core, the actual fiber length will be somewhat longer than the cable length.

The lengths at which the OTDR can be used depend on two things. First is the dynamic range of the units, which sets the minimum and maximum power to the detector. Beyond that, the length is determined by fiber attenuation and losses at splices and connectors. OTDR dynamic range and power losses within the fiber system set the length possible before the backscattered light becomes too weak to be detected. With typical low-loss fibers used in long-distance telecommunications, the OTDR can be used on lengths of 20 to 40 km or more.

A sophisticated OTDR unit offers the technician a great deal of information. The screen in Figure 17-7 shows a loss of 7.65 dB over an 18.44-km link span, for an average of 0.415 dB/km. The large bump at about 13 km represents a reflection, as from a splice

FIGURE 17-7
OTDR screen
(Photo courtesy
of Tektronix,
Inc.)

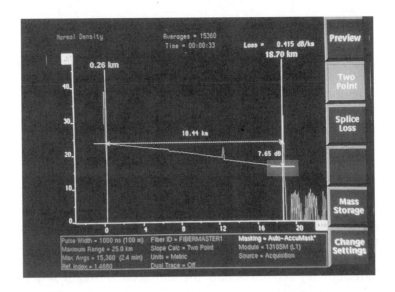

or connector; the small bump at 7 km is another splice, but with minimal loss and very little reflection. Notice also that the bottom of the screen shows conditions of the setup. Two items are of interest. First, source module is 1310SM, meaning the OTDR source is a laser operating at 1310 nm into a single-mode fiber. Second, the refractive index of the fiber is given as 1.4680. The OTDR uses this value to calculate distances based on time. If the refractive index value is wrong, the distances will be somewhat incorrect.

Some OTDRs use plug-in modules to permit operation at 850, 1310, and 1550 nm on both single-mode and multimode fibers. Because the unit uses an internal processor to analyze and display measurements, it can store waveforms. It also allows you to change the resolution from a long view of the overall link span to a closeup of a particular event. You could, for example, zoom in for a close inspection of the reflection in Figure 17-7 to measure the reflection loss.

The OTDR can tell you a great deal about a link and allow you to inspect various sections of a link in detail.

OTDR principles have been adapted to less expensive equipment. An optical fault finder, for example, uses time-domain reflectometry to measure the distance to a fault. Typically, however, it provides only a simple numerical readout of the distance to the fault. The advantage is a low-cost, compact unit, often handheld and battery operated.

Most OTDRs are highly automated and sophisticated, allowing you to zoom into specific areas for a close inspection. Among capabilities are these:

- Zoom in to specific events like connectors, splices, and faults (all of which will show pronounced bumps in the trace) for a closeup look.
- Zoom out to gain an overall picture of the link.

- Measure the total link length.
- Measure distances to specific spots.
- Measure attenuation for the entire link or for specific areas.
- Compare two sections of a link. You can, for example, compare two different interconnections.
- Compare two different links. Many OTDRs have storage capability that will allow you to store a test. This test can then be called up and compared to the current test.
- Save the test results to a computer database.
- Print the test results.

Do you need an OTDR? While link certification can be done with a power meter, nothing beats an OTDR for in-depth troubleshooting. OTDRs have two main drawbacks. First, they are relatively expensive. Second, they require a higher level of operator skill. Even so, OTDRs in recent years have become lower in cost and easier to use. Newer handheld models replace the bulkier lugarounds that usually required a cart.

A useful strategy is to perform certification using power meters, but have one or two OTDRs available for troubleshooting. This can additionally save on training, since only one or two technicians need to specialize in OTDR use.

Some building owners or network administrators, however, require an OTDR plot for every fiber-optic link. This is a peace-of-mind issue, not a performance issue. Compared to a UTP copper link report—detailing NEXT, attenuation, ACR, loop resistance, capacitance, and wire mapping—an optical link report seems short and insubstantial. An OTDR plot adds weight.

USES OF OTDR

OTDR has many uses in research and development and in manufacturing. Here we will discuss three field uses: loss per unit length, splice and connector evaluation, and fault location. These three uses are very important during installation and for system maintenance. Loss per unit length and evaluation of connectors and splices can also be done with an optical power meter. The advantage of OTDR in field applications involving any appreciable length is that only one end of the fiber must be available. The OTDR unit can be placed in a sheltered environment and kept out of manholes. In applications involving short lengths of fibers, such as offices, the power meter might be more convenient.

Loss Per Unit Length

The loss budget of a fiber-optic installation assumes certain fiber attenuation, which is loss per unit length. OTDR can measure attenuation before and after installation. Measurements before installation ensure that all fibers meet specified limits. Measurements after installation check for any increases in attenuation that could result from such things as bends or unexpected loads.

Splice and Connector Evaluation

Here OTDR can be used during the installation processes to ensure that all splices and connector losses are within acceptable limits. After the splice or connector is applied, an OTDR reading checks the loss. If it is unacceptable, another connector or splice is applied.

Some splices are tunable. By rotating one half of the splice in relation to the other, losses from lack of symmetry in the fiber or splice can be minimized. As the splice is rotated, the OTDR display is monitored to determine the point of lowest loss and maximum transmission.

The OTDR unit and operator are usually in a location such as a central telephone office. Communication with the splicer is by radio or telephone.

Fault Location

Faults, such as broken fibers or splices, may occur during or after installation. A telephone line will be broken two or three times during its 30-year lifetime. The location of the fault may not be apparent if the cable is buried or in a conduit or if only the fiber and not the cable is broken. OTDR provides a useful method of locating the fault accurately and thereby saving much time and expense. It certainly beats digging up several kilometers of cable unnecessarily.

TROUBLESHOOTING AN OPTICAL LINK

This section discusses some of the things to check if the link fails.

- Continuity. Use a visual continuity tester—flashlight—to perform a continuity check. This will help identify broken fibers.
- Polarity. Were the fibers crossed during termination? Check end-to-end continuity for each section of the link to discover crossed pairs. Reterminate the cable or reconnect the cable properly.
- Misapplied connector. Check the connector carefully. Check the end finish with a microscope for unacceptable scratches, pits, or other flaws. As appropriate, repolish the connector or reterminate the fiber.
- Dirt. Check the end of the fiber for dirt or films. Clean the connector end with isopropyl alcohol or canned air. Check the core of the coupling adapter for obstructions.

- Make sure the light source and power meter are correctly set and calibrated. For example, using a 850-nm source to test against 1300-nm test conditions significantly distorts results.
- Check the cable plant for tight bends, kinks in the cable, and other unhealthy conditions. These can either increase attenuation or snap fibers in two.
- Make a habit of checking the cable for continuity before installing it. Test it while it is still on the reel. Considering the time it takes to perform this test versus the cost of installing the cable, it is an inexpensive first step in ensuring a quality installation.

FUSION SPLICER

Chapter 11 described a fusion splice. A fusion splicer (Figure 17–8) uses an electric arc to heat fibers to about 2000°C. The fibers melt, are pushed together, and fuse together as they cool.

While fusion splices are used with multimode fibers, single-mode fibers present an additional difficulty because of their core diameter. With multimode fibers, it is sufficient to align the claddings to achieve low splice loss. Single-mode fibers, however, demand that the cores be aligned.

Achieving the best fusion is a complex matter involving not only precise alignment of fibers, but careful application of the correct arc discharge power and timing. An incorrect discharge, for example, can deform the cores. Splicers have adjustable power and time to accommodate different fibers and application needs.

FIGURE 17–8
Fusion splicer
(Photo courtesy
of Siecor
Corporation)

The earliest splicers required a great degree of operator skill to align the fibers, often visually through a microscope. Often power through the fibers was measured as the fibers were being aligned. At the point of maximum power at the junction, the fibers were locked in place and fused. Such power monitoring could be done remotely with an OTDR or locally by techniques that injected a small amount of light into the cladding at a bend in the fiber.

Micropositioners hold the fibers and allow them to be precisely aligned three dimensionally along the x, y, and z axes, either manually or automatically.

Today's splicers have a high degree of computer control and analysis to automate the fusion process. One example—known as profile alignment system—is shown in Figure 17-9. Collimated light is reflected off a mirror, through the fiber at right angles to the axis, and into a video camera. The video camera connects both to a screen for the operator and to a computer that analyzes the images.

Because of the way light is refracted by the different refractive indices of the fiber, the cladding will appear dark and the core will appear light. The splicer's computer analyzes the images to locate the centerlines of the cores. The computer then moves the fibers into alignment. Notice that the camera can move to analyze the fiber on two perpendicular planes. Once the fibers are aligned, the computer will estimate the loss of the splice. If the value is not acceptable, the operator can clean the fibers or recleave them. Once the loss is

FIGURE 17–9

Principle of pro–file alignment of optical fibers (Courtesy of Alcoa Fujikura Ltd.)

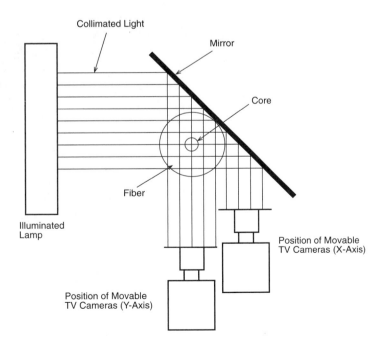

within acceptable limits, the operator initiates the fuse cycle. After the splice is complete, the operator can view the alignment of the cores.

The biggest danger in fusion splicing is the fibers shifting along their axis during fusion due to the surface tension of the molten glass. The splicer, by knowing the parameters of the fiber and the arc discharge, can compensate automatically for axis shift.

POLISHING MACHINE

Polishing machines, such as the device shown in Figure 17-10, help automate fiber polishing. Machine polishing has two main advantages: consistent end finishes and lower termination costs in high-volume production. Consistency in polishing is especially important in single-mode connector applications, where high return loss is required. Even in multimode connector applications, the need for consistency might make the use of machines imperative.

Because polishing machines are a major investment (several thousand dollars), they find widest use in high-volume applications, such as factory production of cable assemblies. In such applications, machine polishing is more economical than hand polishing.

FIGURE 17-10
Fiber-optic polishing machine (Photo courtesy of AMP Incorporated)

In evaluating polishing machines, look for a machine that minimizes any variations in the polishing process. Variations lead to variations in the finish and inconsistent results. Commercial polishing machines are designed to virtually eliminate variations in polishing and to achieve high-quality, consistent polishes. They include the following features:

- *Orbital polishing motion.* An orbital polishing motion, in which the polishing disk rotates in one direction while eccentrically moving in the other direction, creates a counter-rotating polishing pattern. This motion distributes the polishing movement evenly and across a larger area of polishing film than does a simple circular motion.
- *Four-corner holddowns.* Holddown fasteners in all four corners of the connector holder evenly distribute film pressure to minimize off-center polishing. A popular alternative—using center pressure from above—allows the possibility of wiggling or vibrating the connector holder. Such motion increases the vertex offset and leads to inconsistent finishes.
- *Bottom pressure.* The polishing film rests on a rubber polishing pad. These pads, in conjunction with the four-corner holddowns, distribute pressure evenly across the polishing area. Because the pads are resilient, they also help to control the radius of curvature because the ferrule is pressed into the pad during polishing. Different pads (and companion spacers) are available for different styles of connectors. It is important to use the proper pad.
- *Precision connector holders.* Connector holder devices are machined to exacting tolerances so that ferrules are precisely positioned for polishing.

The choice of hand polishing versus machine polishing depends on both the application requirements and on the connector volume. Hand polishing can achieve acceptable results in may applications, particularly those that do not require high return losses. A technician can be trained in a few hours to make acceptable hand polishes. However, consistency is still not as proficient as with a machine polish, although hand-polished connectors perform well within required specifications. For example, a premises-cabling or a fiber-to-the-desk application has generous requirements in relation to typical connector performance.

Some polishing machines are small and light enough for field use, such as wiring a building. They operate either off a battery or another power source.

INSPECTION MICROSCOPE

Portable inspection microscopes (Figure 17-11) allow closeup inspection of cleaved fiber ends and polishing finishes in the field. Closeup inspection is an important quality control procedure for evaluating splices and connectors. It is much simpler than OTDR, although it can be used in conjunction with OTDR.

FIGURE 17-11
Inspection
microscope
(Photo courtesy
of AMP
Incorporated)

The microscope allows both customary through-lens illumination and oblique illumination that reveals pits, scratches, and polishing patterns. Connectors or fibers can be rotated to allow finishes to be viewed across the end, straight on, or at angles in between. Different adapters are required for different connectors and bare fibers. Interchangeable eyepiece lenses and objective lenses allow a magnification range from 40X to 800X.

INSTALLATION KITS

Installation of fiber-optic cables requires many tools, some of which are common and some of which are special to fiber optics and the products being used. An installation kit, such as shown in Figure 17-12, contains all the tools commonly needed to split open multifiber cables, prepare fibers, apply connectors, polish connectors, and inspect the finish.

A partial list of the kit contents includes:

- Screwdriver and nutdriver for preparing organizers
- Side-cutting pliers and utility knife for removing cable sheaths
- Tape measure for measuring cable and fiber stripping lengths
- Cable stripper for removing jacketing
- Fiber strippers for removing buffer coatings
- Scissors for trimming Kevlar strength members
- Roller for mixing epoxy

FIGURE 17-12
Installation kit
(Photo courtesy
of AMP
Incorporated)

- Scribe tool for trimming fibers
- Crimp tool for crimping connectors to fibers
- Polishing materials
- Duster (a canned air for cleaning fibers)
- Heat gun for applying heat-shrink tubing
- Index-matching fluid for lowering or eliminating loss from Fresnel reflections in a fiber-to-fiber interconnection
- Inspection microscope for inspecting fiber end finishes

Such kits provide the proper tools for the job in one convenient location.

SUMMARY

- An optical power meter allows optical power levels to be measured.
- Mode control is often important to repeatable test results.
- An optical time reflectometer is used to evaluate connectors and splices, locate faults, and measure loss per unit length in a fiber.
- A fusion splicer fuses fiber in fiber-to-fiber interconnections.
- A microscope allows end finishes to be inspected.
- A polishing machine produces fine, consistent end finishes.

Review Questions

1. Name the electrical test equipment to which an optical power meter is most easily compared.
2. Describe a method of measuring fiber loss with a power meter.
3. List three methods of obtaining EMD in a fiber under test.
4. For each of the following, say whether the time domain or the frequency domain applies:
   A. Rise time
   B. Bandwidth
   C. Pulse width
   D. 500 MHz
   E. 20 ns
   F. Dispersion
   G. Source spectral width
   H. ODTR
5. Name two factors that limit the distance at which an OTDR unit can be used.
6. List three uses of OTDR.
7. Sketch an OTDR display for a 20-km link with a splice at 5 km and connectors at 10 and 15 km.
8. Besides performing the splice itself, name two other essential activities required of a good splice and that the fusion splicer described in this chapter allows to be performed.
9. How many polishing stages are there in polishing a typical connector? What is the function of each stage?

GLOSSARY

1000Base-LX Gigabit Ethernet over fiber with a long-wavelength (1300-nm) source.

1000Base-SX Gigabit Ethernet over fiber with a short-wavelength (850-nm) source.

1000Base-TX Gigabit Ethernet over Category 5 unshielded twisted-pair cable.

100Base-FX 100-Mbps Ethernet over fiber.

100Base-T4 100-Mbps Ethernet over Category 3 cable.

100Base-TX 100-Mbps Ethernet over Category 5 unshielded twisted-pair cable.

10Base-2 10-Mbps Ethernet over thin coaxial cable.

10Base-5 10-Mbps Ethernet over thick coaxial cable; the original version of Ethernet; not used today.

10Base-F 10-Mbps Ethernet over fiber. Variations include 10Base-FB for the backbone, –FL for point-to-point links, and –FP for passive stars, with 10Base-FL being by far the most common.

10BASE-FB That portion of 10BASE-F that defines the requirements for a fiber backbone.

10BASE-FL That portion of 10BASE-F that defines a fiber-optic link between a concentrator and station.

10Base-T 10Mbps Ethernet over unshielded twisted-pair cable.

4B/5B Encoding A signal modulation scheme in which groups of four bits are encoded and transmitted in five bits in order to guarantee that no more than three consecutive zeroes ever occur; used in FDDI.

802.3 network A 10-Mbps CSMA/CD bus-based network; commonly called Ethernet.

802.5 network A token-passing ring network operating at 4 or 16 Mbps.

8B/10B encoding A signal modulation scheme in which either four bits are encoded into a five-bit word or eight bits are encoded in a 10-bit word to ensure that too many consecutive zeroes do not occur; used in ESCON and Fiber Channel.

Absorption Loss of power in an optical fiber, resulting from conversion of optical power into heat and caused principally by impurities, such as transition metals and hydroxyl ions, and also by exposure to nuclear radiation.

Acceptance angle The half-angle of the cone within which incident light is totally internally reflected by the fiber core. It is equal to arcsin(NA).

AM Amplitude modulation.

Amplitude modulation A transmission technique in which the amplitude of the carrier is varied in accordance with the signal.

Angular misalignment The loss of optical power caused by deviation from optimum alignment of fiber to fiber or fiber to waveguide.

APD   Avalanche photodiode.

Asynchronous transfer mode   A cell-switching network using short, fixed-length cells, operating at a basic speed of 155 Mbps, with other speeds of 25 and 51 Mbps to the desktop and 622 Mbps or higher in the backbone.

ATM   Asynchronous transfer mode.

Attenuation   A general term indicating a decrease in power from one point to another. In optical fibers, it is measured in decibels per kilometer at a specified wavelength.

Attenuation-limited operation   The condition in a fiber-optic link when operation is limited by the power of the received signal (rather than by bandwidth or by distortion.

Avalanche photodiode   A photodiode that exhibits internal amplification of photocurrent through avalanche multiplication of carriers in the junction region.

Backscattering   The return of a portion of scattered light to the input end of a fiber, the scattering of light in the direction opposite to its original propagation.

Bandwidth   A range of frequencies.

Bandwidth-limited operation   The condition in a fiber-optic link when band-width, rather than received optical power, limits performance. This condition is reached when the signal becomes distorted, principally by dispersion, beyond specified limits.

Baseband   A method of communication in which a signal is transmitted at its original frequency without being impressed on a carrier.

Baud   A unit of signaling speed equal to the number of signals symbols per second, which may or may not be equal to the data rate in bits per second.

Beamsplitter   An optical device, such a as a partially reflecting mirror, that splits a beam of light into two or more beams and that can be used in fiber optics for directional couplers.

Bend loss   A form of increased attenuation in a fiber that results from bending a fiber around a restrictive curvature (a macrobend) or from minute distortions in the fiber (microbends).

BER   Bit-error rate.

Bit   A binary digit, the smallest element of information in binary system. A 1 or 0 of binary data.

Bit-error rate   The ratio of incorrectly transmitted bits to correctly transmitted bits.

Broadband   A method of communication in which the signal is transmitted by being impressed on a higher-frequency carrier.

Buffer   A protective layer over the fiber, such as a coating, and inner jacket, or a hard tube.

Buffer coating   A protective layer, such as an acrylic polymer, applied over the fiber, cladding for protective purposes.

Buffer tube   A hard plastic tube, having an inside diameter several times that of a fiber, that holds one or more fibers.

Bus network   A network topology in which all terminals are attached to a transmission medium serving as a bus.

Byte   A unit of 8 bits.

Carrier sense multiple access with collision detection   A technique used to control the transmission channel of a local area network to ensure that there is no conflict between terminals that wish to transmit.

Centro-symmetrical reflective optics   A optical technique in which a concave mirror is used to control coupling of light from one fiber to another.

Channel   A communications path or the signal sent over the channel. Through multiplexing several channels, voice channels can be transmitted over an optical channel.

Cladding   The outer concentric layer that surrounds the fiber core and has a lower index of refraction.

Cladding mode   A mode confined to the cladding; a light ray that propagates in the cladding.

Concentrator   A multiport repeater.

Connector   A device for making connectable/disconnectable connections of a fiber to another fiber, source, detector, or other devices.

Core   The central, light-carrying part of an optical fiber; it has an index of refraction higher than that of the surrounding cladding.

Coupler   A multiport device used to distribute optical power.

CSMA/CD   Carrier sense multiple access with collision detection.

CSR   Centro-symmetrical reflective optics.

Cutoff wavelength   For a single-mode fiber, the wavelength above which the fiber exhibits single-mode operation.

Dark current   The thermally induced current that exists in a photodiode in the absence of incident optical power.

Data rate   The number of bits of information in a transmission system, expressed in bits per second (bps), and which may or may not be equal to the signal or baud rate.

dB   Decibel.

dBm   Decibel referenced to a microwatt.

dBμ   Decibel referenced to a milliwatt.

Decibel   A standard logarithmic unit for the ratio of two powers, voltages, or currents. In fiber optics, the ratio is power.

$$dB = 10 \log_{10}\left(\frac{P_1}{P_2}\right)$$

Dense wavelength-division multiplexing   A form of wavelength-division multiplexing in which the different wavelengths being multiplexed are in the same optical window. Typical wavelength separations are 100 or 200 GHz, or about 0.8 or 1.6 nm.

Detector   An optoelectronic transducer used in fiber optics for converting optical power to electric current. In fiber optics, usually a photodiode.

DFB   Distributed-feedback laser.

Diameter-mismatch loss   The loss of power at a joint that occurs when the transmitting half has a diameter greater than the diameter of the receiving half. The loss occurs when coupling light from a source to fiber, from fiber to fiber, or from fiber to detector.

Dichroic filter   An optical filter that transmits light selectively according to wavelength.

Diffraction grating   An array of fine, parallel, equally spaced reflecting or transmitting lines that mutually enhance the effects of diffraction to concentrate the diffracted light in a few directions determined by the spacing of the lines and by the wavelength of the light.

Dispersion   A general term for those phenomena that cause a broadening or spreading of light as it propagates through an optical fiber. The three types are modal. material, and waveguide.

Distortion-limited operation   Generally synonymous with bandwidth-limited

Distributed-feedback laser   A laser having a built-in mechanism to suppress all but a very narrow band of emitted wavelengths. DFBs are characterized by very narrow line widths and high operating speeds.

Duplex cable   A two-fiber cable suitable for duplex transmission.

Duplex transmission   Transmission in both directions, either one direction at a time (half duplex) or both directions simultaneously (full duplex).

Duty cycle   In a digital transmission, the ratio of high levels to low levels.

DWDM   Dense wavelength-division multiplexing.

EDFA   Erbium-doped fiber amplifier.

Electromagnetic interference   Any electrical or electromagnetic energy that causes undesirable response, degradation , or failure in electronic equipment. Optical fibers neither emit nor receive EMI.

EMD   Equilibrium mode distribution.

EMI   Electromagnetic interference.

Equilibrium mode distribution   The steady modal state of a multimode fiber in which the relative power distribution among modes is independent of fiber length.

Erbium-doped fiber amplifier   A type of fiber that amplifies 1550-nm optical signals when pumped with a 980- or 1480-nm light source.

ESCON   An IBM channel control system based on fiber optics.

Excess loss   In a fiber-optic coupler, the optical loss from that portion of light that does not emerge from the nominally operation ports of the device.

Extrinsic loss   In a fiber interconnection, that portion of loss that is not intrinsic to the fiber but is related to imperfect joining, which may be caused by the connector or splice.

Eye Pattern   A qualitative method of evaluating signals by superimposing a long stream of bits on one another on an oscilloscope so that rise times, fall times, pulse width, and noise margins are all displayed visually.

Fall time   The time required for the trailing edge of a pulse to fall from 90% to 10% of its amplitude; the time required for a component to produce such a result. "Turn off time." Sometimes measured between the 80% and 20% points.

FDDI   Fiber Distributed Data Interface.

Fiber Bragg grating   A fiber whose refractive index has been altered periodically so that the fiber acts as an optical filter.

Fiber channel   An industry-standard specification for computer channel communications over fiber optics and offering transmission speeds from 132 Mbaud to 1062 Mbaud and transmission distances from 1 to 10 km.

Fiber distributed datainterface network   A token-passing ring network designed specifically for fiber optics and featuring dual counterrotatings rings and 100 Mbps operation.

Fiber-optic interrepeater link   A standard defining a fiber-optic link between two repeaters in a IEEE 802.3 network.

FM   Frequency modulation.

FOIRL   Fiber-optic interreapeater link.

Frequency modulation   A method of transmission in which the carrier frequency varies in accordance with the signal.

Fresnel reflection   The reflection that occurs at the planar junction of two materials having different refractive indices; Fresnel reflection is not a function of the angle of incidence.

Fresnel reflection loss   Loss of optical power due to Fresnel reflections.

Fused coupler   A method of making a multimode or single-mode coupler by wrapping fibers together, heating them, and pulling them to form a central unified mass so that light on any input fiber is coupled to all output fibers.

Gap loss   Loss resulting from the end separation of two axially aligned fibers.

Graded-index fiber   An optical fiber whose core has a nonuniform index of refraction. The core is composed of concentric rings of glass whose refractive indices decrease from the center axis. The purpose is to reduce modal dispersion and thereby increase fiber bandwidth.

Group-loop noise   Noise that results when equipment is grounded at ground points having different potentials and thereby creating an unintended current path. The dielectric of optical fibers provides isolation that eliminates ground loops.

HFC   Hybrid fiber-coax.

Hybrid fiber-coax   A system, used commonly in cable television, using fiber in the backbone and coaxial cable for local distribution to the home.

IDP   Integrated detector/preamplifier.

Index of refraction   The ratio of the velocity of light in free space to the velocity of light in a given material. Symbolized by n.

Index-matching material   A material, used at optical interconnection, having a refractive index close to that of the fiber core and used to reduce Fresnel reflections.

Insertion loss   The loss of power that results from inserting a component, such as a connector or splice, into a previously continuous path.

Integrated detector/preamplifier   A detector package containing a pin photodiode and transimpedance amplifier.

ISO   International Standard Organization.

Jitter   The variation of a pulse from its ideal position in time.

LAN   Local area network.

Laser   A light source producing, through stimulated emission, coherent, near monochromatic light. Lasers in fiber optics are usually solid-state semiconductor types.

Lateral displacement loss   The loss of power that results from lateral displacement from optimum alignment between two fibers or between a fiber and an active device.

LED   Light-emitting diode.

Light-emitting diode   A semiconductor diode that spontaneously emits light from the pn junction when forward current is applied.

Local area network   A geographically limited network interconnecting electronic equipment.

Material dispersion   Dispersion resulting from the different velocities of each wavelength in an optical fiber.

Media converter   A device whose main purpose is to convert signals from electrical to optical and vice versa; used, for example, to allow fiber runs in an otherwise copper-based network.

MFD   Mode field diameter.

Misalignment loss   The loss of power resulting from angular misalignment, lateral displacement, and end separation.

Modal dispersion   Dispersion resulting from the different transit lengths of different propagating modes in a multimode optical fiber.

Mode   In guided-wave propagation, such as through a waveguide or optical fiber, a distribution of electromagnetic energy that satisfies Maxwell's equations and boundary conditions. Loosely, a possible path followed by light rays.

Mode coupling   The transfer of energy between modes. In a fiber, mode coupling occurs until EMD is reached.

Mode field diameter   The diameter of optical energy in a single-mode fiber. Because the MFD is greater than the core diameter, MFD replaces core diameter as a practical parameter.

Mode filter   A device used to remove high-order modes from a fiber and thereby simulate EMD.

Modulation   The process by which the characteristic of one wave (the carrier) is modified by another wave (the signal). Examples include amplitude modulation (AM), frequency modulation (FM), and pulse-coded modulation(PCM).

MQW   Multiquantum well.

Multimode fiber A type of optical fiber that supports more than one propagating mode.

Multiplexing The process by which two or more signals are transmitted over a single communications channel. Examples include time-division multiplexing and wavelength-division multiplexing.

Multiquantum well A method of building the active lasing region of a laser, using a complex structure to guide and contain photons.

NA Numerical aperture.

NA–mismatch loss The loss of power at a joint that occurs when the transmitting half has an NA greater than the NA of the receiving half. The loss occurs when coupling light from a source to fiber, from fiber to fiber, or from fiber to detector.

Nonzero dispersion-shifted fiber A fiber whose properties are dispersion shifted to a region other than the point of zero dispersion.

Numerical aperture The "light-gathering ability" of a fiber, defining the maximum angle to the fiber axis at which light will be accepted and propagated through the fiber. NA = sin 0, where 0 is the acceptable angle. NA is also used to describe the angular spread of light from a central axis, as in exiting a fiber, emitting from a source, or entering a detector.

OADM Optical add-drop multiplexer.

OFA Optical-fiber amplifier.

Open Standard Interconnect A seven-layer model defined by ISO for defining a communication network.

Optical add-drop multiplexer A device that allows channels to be added to or removed from a multiplexed optical transmission.

Optical time-domain reflectometry A method of evaluating optical fibers based on detecting backscattered (reflected) light. Used to measure fiber attenuation, evaluate splice and connector joints, and locate faults.

Optical-fiber amplifier A fiber-based signal amplifier, such as an erbium-doped fiber, which amplifies an optical signal without requiring it to be first converted to an electrical signal.

OSI Open Standards Interconnect.

OTDR Optical time-domain reflectometry.

PC Physical contact.

PCM Pulse-coded modulation.

PCS Plastic-clad silica.

Photodetector An optoelectronic transducer, such as a pin photodiode or avalanche photodiode.

Photodiode A semiconductor diode that produces current in response to incident optical power and used as a detector in fiber optics.

Photon A quantum of electromagnetic energy. A "particle" of light.

Physical contact connector   A connector designed with a radiused tip to ensure physical contact of the fibers and thereby increase return reflection loss.

Pigtail   A short length of fiber permanently attached to a component, such as a source, detector, or coupler.

Pin photodiode   A photodiode having a large intrinsic layer sandwiched between p-type and n-type layers.

Pistoning   The movement of a fiber axially in and out of a ferrule end, often caused by changes in temperature.

Plastic fiber   An optical fiber having a plastic core and plastic cladding.

Plastic–clad silica fiber   An optical fiber having a glass core and plastic cladding.

Plenum   The air space between walls, under structural floors, and above drop ceilings, which can be used to route intrabuilding cabling.

Plenum cable   A cable whose flammability and smoke characteristics allow it to be routed in a plenum area without being enclosed in a conduit.

POF   Plastic optical fiber.

Pulse spreading   The dispersion of an optical signal with time as it propagates through an optical fiber.

Pulse-coded modulation   A technique in which an analog signal, such as a voice, is converted into a digital signal by sampling the signal's amplitude and expressing the different amplitudes as a binary number. The sampling rate must be twice the highest frequency in the signal.

Quantum efficiency   In a photodiode, the ratio of primary carriers (electron-hole pairs) created to incident photons. A quantum efficiency of 70% means 7 out of 10 incident photons create a carrier.

Rayleigh scattering   The scattering of light that results from small inhomogeneities in material density or composition.

Regenerative repeater   As repeater designed for digital transmission that both amplifies and reshapes the signal.

Repeater   A device that receives, amplifies (and perhaps reshapes), and retransmits a signal. It is used to boost signal levels when the distance between repeaters is so great that the received signal would otherwise be too attenuated to be properly received.

Responsivity   The ratio of a photodetector's electrical output to its optical input is amperes/watt.

Return reflection   Reflected optical energy that propagates backward to the source in an optical fiber.

Return reflection loss   The attenuation of reflected light; high return loss is desirable, especially in single-mode fibers.

Ring network   A network topology in which terminals are connected in a point-to-point serial fashion in an unbroken circular configuration.

Rise time   The time required for the leading edge of a pulse to rise from 10% to 90% of its amplitude; the time required for a component to produce such a result. "Turn-on time." Sometimes measured between the 20% and 80% points.

SAN   Storage-area network.

Sensitivity   For a fiber-optic receiver, the minimum optical power required to achieve a specified level of performance, such as BER.

Shot noise   Noise caused by random current fluctuations arising from the discrete nature of electrons.

Signal-to-noise ratio   The ratio of signal power to noise power.

Simplex cable   A term sometimes used for a single-fiber cable.

Simplex transmission   Transmission in one direction only.

Single-mode fiber   An optical fiber that supports only one mode or light propagation above the cutoff wavelength.

SNR   Signal-to-noise ratio.

Soliton   An optical pulse that does not disperse over distance.

Sonet   Synchronous optical network, an international standard for fiber-optic digital telephony.

Source   The light emitter, either an LED or laser diode, in a fiber-optic link.

Spectral width   A measure of the extent of a spectrum. For a source the width of wavelengths contained in the output at one half of the wavelength of peak power. Typical spectral widths are 20 to 60 nm for an LED and 2 to 5 nm for a laser diode.

Splice   An interconnection method for joining the ends of two optical fibers in a permanent or semipermanent fashion.

Star coupler   A fiber-optic coupler in which power at any input port is distributed to all output ports.

Star network   A network in which all terminals are connected through a single point, such as a star coupler.

Steady state   Equilibrium mode distribution.

Step-index fiber   An optical fiber, either multimode or single mode, in which the core refractive index is uniform throughout so that a sharp step in refractive index occurs at the core-to-cladding interface. It usually refers to a multimode fiber.

Storage-area network   A network aimed principally at connecting servers and storage devices in "backend" applications.

Strength member   That part of a fiber-optic cable composed of Kevlar aramid yarn, steel strands, or fiberglass filaments that increase the tensile strength of the cable.

Tap loss   In a fiber-optic coupler, the ratio of power at the tap port of the power at the input port.

Tap port   In a coupler in wish the splitting ratio between output ports is not equal, the output port containing the lesser power.

TDM   Time-division multiplexing.

Tee coupler   A three-port optical coupler.

VCSEL   Vertical-cavity surface-emitting laser.

Vertical-cavity surface-emitting laser   A low-cost laser that emits light from the surface rather from the edge.

INDEX